A fishery manager's guidebook

Management measures and their application

Edited by
Kevern L. Cochrane
Senior Fishery Resources Officer
Fishery Resources Division
FAO Fisheries Department

Rome, 2002

The designations employed and the presentation of the material in
this information product do not imply the expression of any opinion
whatsoever on the part of the Food and Agriculture Organization
of the United Nations concerning the legal status of any country,
territory, city or area or of its authorities, or concerning the
delimitation of its frontiers or boundaries.

ISBN 92-5-104773-1

All rights reserved. Reproduction and dissemination of material in this
information product for educational or other non-commercial purposes are
authorized without any prior written permission from the copyright holders
provided the source is fully acknowledged. Reproduction of material in this
information product for resale or other commercial purposes is prohibited
without written permission of the copyright holders. Applications for such
permission should be addressed to the Chief, Publishing and Multimedia Service,
Information Division, FAO, Viale delle Terme di Caracalla, 00100 Rome, Italy or
by e-mail to copyright@fao.org

© **FAO 2002**

PREPARATION OF THIS DOCUMENT

This document was prepared to supplement the available information on implementation of the Code of Conduct for Responsible Fisheries, in particular to build on that contained in the FAO Technical Guidelines for Responsible Fisheries No. **4**: Fisheries Management. The different chapters were prepared by experts in their fields working under contract to FAO.

The initial versions of all chapters were reviewed by Messrs Les Clark, Jean-Jacques Maguire and Patrick McConney, who are all thanked for careful and constructive comments which resulted in considerable improvements to the Guidebook as a whole. Thanks are also due to Alain Bonzon, Jorge Csirke, George Everett, Serge Garcia, Andy Smith, Joel Prado, Annick Van Houtte and Rolf Willmann of FAO for helpful input on various chapters. Anne Van Lierde's substantial assistance in editing and the layout of this FAO Fisheries Technical Paper is gratefully acknowledged. The cover illustration was designed by Emanuela D'Antoni.

Distribution:

FAO Fisheries Department
FAO Regional and Subregional Fishery Officers
Directors of Fisheries
COFI selected addresses

Cochrane, K.L. (ed.)
A fishery manager's guidebook. Management measures and their application.
FAO Fisheries Technical Paper. No. 424. Rome, FAO. 2002. 231p.

ABSTRACT

This publication was prepared to promote and to provide support in the implementation of the Code of Conduct for Responsible Fisheries, especially Article 7: Fisheries Management. As such it is also intended to supplement the FAO Technical Guidelines for Responsible Fisheries No. 4: Fisheries Management. It is intended primarily for the practising fishery manager and decision-maker, with particular emphasis on developing countries, although it is hoped that the volume will also be of interest to managers in developed countries.

Fisheries management is a complex and evolving discipline and much is still being learnt about what it involves, what works and what doesn't. The problem is compounded by the fact that fisheries management as a coherent discipline is still poorly defined and frequently equally poorly understood. This publication strives to identify the primary tasks in management of capture fisheries, with particular emphasis on sustainable utilization of the biological resources, and to demonstrate how these tasks should be undertaken in an integrated and coordinated manner to obtain the desired benefits from the biological resources in a sustainable and responsible manner.

The Guidebook is divided into nine different chapters, individually authored by experts in the field from around the world. Chapter 1 provides an introduction to fisheries management, discussing what it is and who or what is the fishery manager. It discusses the inter-relationships between goals and objectives and management plans, measures and strategies, and examines some of the primary issues which need to be considered by the managers in executing their task.

Chapter 2 provides an overview of the different types of fishing gear used in fisheries and the impacts each can have on the target species, bycatch species and rest of the ecosystem and discusses how management can regulate the use and characteristics of fishing gear. Chapter 3 examines the role of closed areas and closed seasons in fisheries management, considering the different goals they can serve and their potential advantages and disadvantages. Consideration of a number of case studies demonstrates their use in practice. Chapter 4 examines direct input (effort) and output (catch) control in fisheries and provides insight into different types of input and output control, the structures and capacity needed for their application, and some of the more common problems that can be encountered in using them. All of these management measures and strategies should be developed in order to achieve the agreed objectives for the fishery, and Chapter 5 describes how the fisheries manager can determine the most appropriate strategies for their objectives, with particular emphasis on the role and use of scientific and other information on the fishery and the resources it exploits.

It is now generally accepted that open access fisheries are biologically, economically and socially damaging and Chapter 6 examines the topical question of allocating use rights in fisheries. It discusses the nature and forms of use rights and how they are implemented. It demonstrates that use rights are also a form of management measure and different systems of use rights will assist in achieving different objectives. Chapter 7 continues the emphasis on the user groups or interested parties and considers the importance of involving these parties as partners in fisheries management. It presents the different scopes and scales which such partnerships can cover, and examines the benefits and problems in managing in partnership, including the conditions required for effective partnerships.

Finally, Chapter 9 describes the importance of formulating management plans and what should be included in fisheries management plans. It examines their implementation and the need to review them periodically. It also provides some case studies of their development and role in a range of fishery types.

TABLE OF CONTENTS

	Page
Chapter 1. FISHERIES MANAGEMENT by K.L. Cochrane	1
Chapter 2. THE USE OF TECHNICAL MEASURES IN RESPONSIBLE FISHERIES: REGULATION OF FISHING GEAR by A. Bjordal	21
Chapter 3. THE USE OF TECHNICAL MEASURES IN RESPONSIBLE FISHERIES: AREA AND TIME RESTRICTIONS by S. Hall	49
Chapter 4. INPUT AND OUTPUT CONTROLS: THE PRACTICE OF FISHING EFFORT AND CATCH MANAGEMENT IN RESPONSIBLE FISHERIES by J. Pope	75
Chapter 5. THE USE OF SCIENTIFIC INFORMATION IN THE DESIGN OF MANAGEMENT STRATEGIES by K.L. Cochrane	95
Chapter 6. USE RIGHTS AND RESPONSIBLE FISHERIES: LIMITING ACCESS AND HARVESTING THROUGH RIGHTS-BASED MANAGEMENT by A.T. Charles	131
Chapter 7. PARTNERSHIPS IN MANAGEMENT by E. Pinkerton	159
Chapter 8. FISHERY MONITORING, CONTROL AND SURVEILLANCE by P.E. Bergh and S. Davies	175
Chapter 9. DESIGN AND IMPLEMENTATION OF MANAGEMENT PLANS by D. Die	205
Glossary	221
Authors' addresses and short biographies	227

CHAPTER 1
FISHERIES MANAGEMENT

by

Kevern L. COCHRANE

Fisheries Department, FAO

1. WHY DO WE NEED THIS GUIDEBOOK? 1
2. WHAT IS FISHERIES MANAGEMENT? 3
3. THE WORKING PRINCIPLES OF FISHERIES MANAGEMENT 4
4. WHO IS THE FISHERY MANAGER? 6
5. WHAT CONSTITUTES A MANAGEMENT AUTHORITY? 7
6. GOALS AND OBJECTIVES: WHO NEEDS THEM IN A FISHERY? 8
7. MANAGEMENT PLANS, MEASURES AND STRATEGIES 10
8. PRIMARY CONSIDERATIONS IN FISHERIES MANAGEMENT 11
    8.1 Biological Considerations 12
    8.2 Ecological and Environmental Considerations 12
    8.3 Technological Considerations 13
    8.4 Social and Cultural Considerations 14
    8.5 Economic Considerations 15
    8.6 Considerations Imposed by Other Parties 16
9. IS ALL THIS LEGAL? 16
    9.1 National Legislation 16
    9.2 International legislation and instruments 17
10. RECOMMENDED READING 18

1. WHY DO WE NEED THIS GUIDEBOOK?

Fish and fisheries are an integral part of most societies and make important contributions to economic and social health and well-being in many countries and areas. It has been estimated that approximately 12.5 million people are employed in fishery-related activities, and in recent years global production from capture fisheries has tended to vary between approximately 85 and 90 million tonnes. The products from these fisheries are used in a wide variety of ways, ranging from subsistence use to international trade as highly sought-after and highly-valued items. The value of fish traded internationally is approximately US$40 billion per year.

Despite this enormous importance and value, or more correctly, because of these attributes, the world's fish resources are suffering the combined effects of heavy exploitation and, in some cases, environmental degradation. The FAO (2000) estimated that, in 1999, 47% of the 441 stocks for which some information on status was available were fully exploited, 18% overexploited, 9% depleted and 1% recovering. This pattern is broadly consistent with similar statistics available from other regions. For example, the National Marine Fisheries Service of the United States of America estimated in 1998 that 30% of the stocks in the waters of that country for which information was available were overfished. In the waters of the European Community, it was estimated that in 1990, 57% of the stocks were 'heavily exploited'. Fish stocks throughout the rest of the globe are likely to be in a similar condition to those in these regions.

There are many reasons for this unacceptable state of affairs, but the primary reasons all come down to a failure in fisheries governance in most countries. The responsibility for declining stocks and falling economic returns and employment opportunities in fisheries must be shared amongst fishers, fisheries management authorities, fishery scientists and those involved in environmental degradation. Not all of the underlying problems lie within the realm of fisheries management, but the fisheries manager is the person who is most often in the best position to observe and record what is happening in the fisheries under his or her jurisdiction, to establish the underlying cause or causes of any problems, to rectify those within their jurisdiction, and to bring the others to the attention of both the interested parties in fisheries and those with a responsibility covering the external causes. However, all too often the fisheries manager remains either unaware of the state of the resources, or fails to act sufficiently as the fisheries slip further and further into decay and crisis, or both. This is rarely, if ever, a deliberate choice and more often comes down to a lack of available information, an incomplete understanding of the nature of the task of fisheries management, and inadequate resources, structures and support to address the problems and utilise the resources in a planned and efficient manner.

The FAO Code of Conduct for Responsible Fisheries was produced in response to global concern over the clear signs of over-exploitation of fish stocks throughout the world and to recommend new approaches to fisheries management which included conservation, environmental, social and economic considerations. It was developed by and through FAO and was accepted as a voluntary instrument by the 28th Session of the FAO Conference in October 1995. In addition to five introductory Articles and one on General Principles, the Code contains six thematic articles on Fisheries Management, Fishing Operations, Aquaculture Development, Integration of Fisheries into Coastal Area Management, Post-Harvest Practices and Trade, and Fisheries Research. Overall it incorporates the key considerations in responsible fisheries and provides guidance on how these should be addressed in order to ensure sustainable and responsible fisheries. Subsequently, FAO has produced a number of Technical Guidelines on different aspects of the Code, including the FAO Technical Guidelines for Responsible Fisheries No. 4: Fisheries Management, which specifically addresses Article 7: Fisheries Management of the Code. The following Technical Guidelines had been produced at the time of printing this Guidebook (late 2001):

- No. 1 Fishing Operations
- No. 1, Suppl.1 Fishing Operations. Vessel Monitoring System
- No. 2 Precautionary Approach to Capture Fisheries and Species Introductions
- No. 3 Integration of Fisheries into Coastal Area Management
- No. 4 Fisheries Management
- No. 4, Suppl.1 Fisheries Management. Conservation and Management of Sharks
- No. 5 Aquaculture Development
- No. 5, Suppl.1 Aquaculture Development. Good Aquaculture Feed Manufacturing Practice
- No. 6 Inland Fisheries

No. 7 Responsible Fish Utilization

No. 8 Indicators for Sustainable Development of Marine Capture Fisheries

This Guidebook has been produced to supplement the Code of Conduct and the Technical Guidelines No. 4 (FAO, 1997), in order to provide managers with additional and more detailed information in determining the scope of their task and how to execute their fisheries management functions. It cannot address in detail all of the issues which fall under Article 7 of the Code, the Article dealing directly with fisheries management, but focuses mainly on those aspects directly related to strategic and operational management of the fisheries themselves and the resources on which they depend. These are the areas in which the fisheries manager generally holds a direct mandate and responsibility.

2. WHAT IS FISHERIES MANAGEMENT?

There is no clear and generally accepted definition of fisheries management. We do not wish to get embroiled in a debate about exactly what fisheries management is and isn't, but use here the working definition used in the Technical Guidelines to provide a summary of the task of fisheries management:

> "The integrated process of information gathering, analysis, planning, consultation, decision-making, allocation of resources and formulation and implementation, with enforcement as necessary, of regulations or rules which govern fisheries activities in order to ensure the continued productivity of the resources and the accomplishment of other fisheries objectives.

From this description, it can be seen that fisheries management involves a complex and wide-ranging set of tasks, which collectively have the achievement of sustained optimal benefits from the resources as the underlying goal. These tasks are also summarised in Figure 1.

There has been a lot of interest in recent years in moving from fisheries management focused essentially on single-species or single fisheries, to management with an ecosystem orientation. This expanded approach has been termed ecosystem-based fisheries management (EBFM) and was recently discussed at "The Reykjavik Conference on Responsible Fisheries in the Marine Ecosystem" (1-4 October 2001), which was organised jointly by FAO and the Governments of Iceland and Norway. The Conference agreed on the Reykjavik Declaration[1] which included an affirmation "that incorporation of ecosystem considerations implies more effective conservation of the ecosystem and sustainable use" and also a reaffirmation of the principles of the FAO Code of Conduct for Responsible Fisheries.

In writing this Guidebook, the authors have implicitly accepted EBFM as a principle inherent in fisheries management and, while the term is not explicitly referred to in the Guidebook, its principles and requirements in fisheries management are incorporated and discussed throughout the volume.

---

[1] http://www.refisheries2001.org/

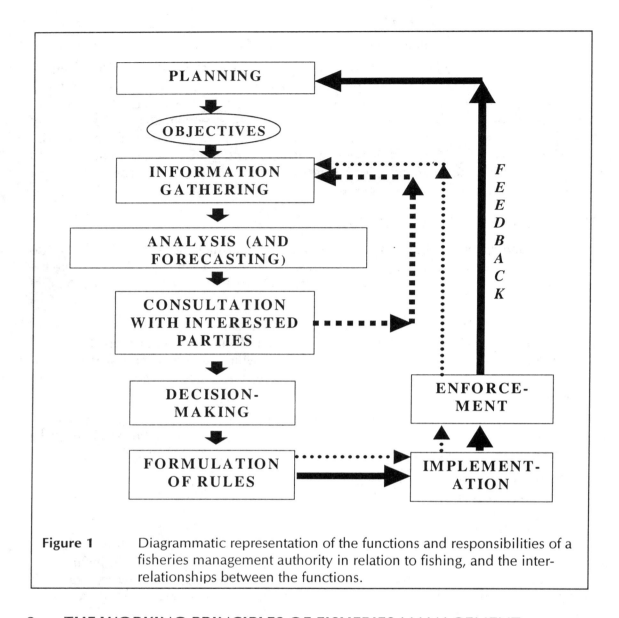

Figure 1 Diagrammatic representation of the functions and responsibilities of a fisheries management authority in relation to fishing, and the inter-relationships between the functions.

3. THE WORKING PRINCIPLES OF FISHERIES MANAGEMENT

The above description presents a complex and possibly confusing picture of all the tasks which need to be considered by the fisheries manager. Some of this complexity can be reduced by attempting to highlight the underlying key issues. There are both benefits and risks in attempting to simplify a subject and over-simplification can lead to neglect of important details. However, simplification can facilitate understanding important principles and highlighting the broad areas which need attention. Arising from the considerations discussed above, a number of key principles can be identified which may serve to focus attention on the starting points for effective fisheries management (Table 1).

Table 1 Suggested fundamental principles of fisheries management (modified from Cochrane, 2000).

	Principle	Management Function	Pertinent Chapters
1.	Fish stocks and communities are finite and biological production constrains the potential yield from a fishery.	The potential yield needs to be estimated and the biological constraints identified.	1, 5
2.	i) Biological production of a stock is a function of the size of the stock and ii) it is also a function of the ecological environment. It is influenced by natural or human-induced changes in the environment.	i) Target reference points need to be established through data collection and fisheries assessment; and ii) environmental impacts should be identified and monitored, and the management strategy adjusted in response as necessary.	1, 5
3.	Human consumptive demands on fish resources are fundamentally in conflict with the constraint of maintaining a suitably low risk to the resource. Further, modern technology provides humans with the means, and demand for its benefits provides the motivation, to extract fish biomass at rates much higher than it can be produced.	Realistic goals and objectives must be set. Achieving the objectives will inevitably require controls on fishing effort and capacity.	2 – 5, 9 5, 6, 8
4.	In a multispecies fishery, which description encompasses almost all fisheries, it is impossible to maximise or optimise the yield from all fisheries simultaneously.	Realistic goals and objectives must be established across ecosystems, so as to manage for species and fisheries interactions.	2 - 5
5.	Uncertainty pervades fisheries management and hinders informed decision-making. The greater the uncertainty, the more conservative should be the approach (i.e. as uncertainty increases, realised yield as a proportion of estimated maximum average yield should be decreased).	Risk assessment and management must be done in development and implementation of management plans, measures and strategies.	5, 9
6.	The short-term dependency of society on a fishery will determine the relative priority of the social and/or economic goals in relation to sustainable utilisation.	Fisheries cannot be managed in isolation and must be integrated into coastal zone and fisheries policy and planning and national policies.	5-7, 9

Table 1 (cont.) Suggested fundamental principles of fisheries management (modified from Cochrane, 2000).

	Principle	Management Function	Pertinent Chapters
7.	A sense of ownership and a long-term stake in the resource for those (individuals, communities or groups) with access are most conducive to maintaining responsible fisheries.	A system of effective and appropriate access rights must be established and enforced.	6-8
8.	Genuine participation in the management process by fully-informed users is consistent with the democratic principle, facilitates identification of acceptable management systems and encourages compliance with laws and regulations	Communication, consultation and co-management should underlie all stages of management	7-9

In keeping with the integrated nature of fisheries ecosystems, these principles cannot be considered in isolation in considering how best to manage fisheries: their implications and consequences overlap, complement and confound each other which is what makes fisheries management so demanding and challenging. Nevertheless, the consequences of the principles for fisheries give rise to the fundamental nature and tasks of fisheries management, and hence to the general structure of this Guidebook (Table 1).

4. WHO IS THE FISHERY MANAGER?

The Technical Guidelines (FAO, 1997) suggest that fisheries management institutions have two major components: the fisheries management authority and the interested parties. The fishers and fishing companies would usually be the major participants amongst the interested parties. The fisheries management authority is that entity which has been given the mandate by the State (or States in the case of an international authority) to perform specific management functions. In many countries that authority would be a Department of Fisheries or, within a broader Department, a Division of Fisheries. However, a fisheries management authority does not have to fall directly within central government, and could be, for example, provincial, local, parastatal or private. Any one of these arrangements can function effectively, given an adequate legal framework in which to operate and the resources necessary to fulfil their function.

Who, then, within this authority is the fisheries manager and to whom is this Guidebook addressed? In fact, despite the fact that we have deliberately used the term in the title, we suggest that in modern fisheries management, there is rarely a single individual who fulfils the functions of "fisheries manager". The head of the authority, for example, a Director of Fisheries, may have overall responsibility for implementing fisheries management and, as well as being accountable and responsible for the advice passed on from his or her Department to the political decision-maker, may act in an overall coordinating role. However, this individual is unlikely to, and generally should not, have sole responsibility for receiving information, formulating advice and making and implementing decisions. Fisheries management is a complex and multi-faceted discipline and requires input from a range of perspectives. It is therefore inappropriate to expect any individual to fulfil this function on their own. In addition, as discussed above and reflected in Paragraph 7.1.2 of the Code of Conduct, fisheries management should involve the legitimate interested parties in the management process.

Perhaps the closest we can come to a "fisheries manager" is the management authority as a whole, including technical experts, monitoring, control and surveillance (MCS) units, administrative units, the executive body of the formal authority, the consultative mechanisms, the advisory body where one exists, and the responsible political head who is often a Minister. Each member of these functional bodies is, to some extent, a fishery manager and this Guidebook is aimed at all of them. It is not designed to go into great technical and operational detail on each function or task as this would require a set of Guidebooks. Instead, it is intended to give a holistic picture of how the different functions should interact in a fisheries management authority in order to develop appropriate objectives, management strategies and plans, and how to encourage all those participating in a fishery to collaborate in and adhere to the agreed strategy.

5. WHAT CONSTITUTES A MANAGEMENT AUTHORITY?

The responsibility for fisheries management rests with the designated fisheries arrangement or organization which, in this Guidebook, we have referred to without distinction as the fisheries management authority. Following the practice used in the Technical Guidelines on Fisheries Management (FAO, 1997) the term is used broadly here to describe that legal entity which has been designated by the State as having the mandate to perform specified fisheries management functions. In practice, it may be a national or provincial ministry, a department within a ministry, or an agency and could be governmental, parastatal or private. In the case of shared resources it should be international.

The area of competence, geographical area, fish resources and fisheries for which a given management authority is responsible must be precisely specified in each case in the appropriate legislation. The task of an authority is diverse and complex and as a result, fisheries management authorities are normally divided into institutional support structures: the fisheries management institutions. The institutions need to encompass the basic tasks and functions of fisheries management as described in Section 2 and Figure 1 of this Chapter. The actual institutional structure and mechanisms may differ from authority to authority and it would be inappropriate for us in this Guidebook to attempt to prescribe any specific set of characteristics as representing the 'best' institutional structure and processes. What is best in each case will depend in large part on the specific circumstances and context. What is universal, however, is that it is essential for the different institutions concerned with management of any fishery or fisheries to be able to interact effectively, requiring good channels of communication and feedback. The institutions must also be seen by the different interested parties as being legitimate.

The need for collaboration between the authority and the interested parties is as important as collaboration between the institutions within the authority and is discussed at length in Chapter 7. That chapter also examines the pre-requisites for effective partnerships between the management authority and the interested parties and the different types of partnership which can be considered.

It is common and frequently desirable for the national government to devolve all or some fisheries management functions to local government or to smaller groups such as fishing communities. In such cases, it is essential to specify precisely the responsibilities and functions, including the geographical area, falling under this local authority or smaller group. The institutions within the local authority must follow the same principles as those for a national authority as discussed here.

The Code of Conduct requires that fisheries management should be concerned with the whole stock over its entire area of distribution (Code of Conduct, Paragraph 7.3.1) and therefore that States should cooperate in the management of transboundary, straddling, highly migratory and high seas fish stocks exploited by two or more states (Paragraph 7.1.3). General rules for cooperation towards conservation of such fish stocks are foreseen in the United Nations

Convention on the Law of the Sea of 10 December 1982, and in the 1995 UN Fish Stocks Agreement (see Table 2). The responsibilities, functions and structure of international or regional fisheries authorities will usually not differ substantively from those of national authorities.

6. GOALS AND OBJECTIVES: WHO NEEDS THEM IN A FISHERY?

The over-riding goal of fisheries management is the long-term sustainable use of the fisheries resources (Code of Conduct, Paragraph 7.2.1). Achieving this requires a proactive approach and should involve actively seeking ways to optimise the benefits derived from the resources available. This rarely happens, though, and fisheries management is still most commonly practised as a reactive activity, where decisions are made and actions taken largely in response to problems or crises. The resulting crisis decisions are then normally attempts merely to solve the immediate problems without properly considering the broader perspective and the longer-term objectives. Such an approach may succeed in maintaining dissatisfaction sufficiently low to avoid major conflict, but it is extremely unlikely to result in the best use of the marine resources being exploited by the fishery.

The first step in proactive fisheries management is to decide what is meant by optimising the benefits for each fishery - what can the State or the collection of legitimate interested parties agree on as being optimal benefits? This may be described in general terms in the national fisheries policy which must be the starting point for determining the specific objectives for each fishery. The broad goals stated in the fisheries policy may need to be tailored for a specific fishery, but the goals for each fishery should be consistent with the policy.

In general terms, the goals in fisheries management can be divided into four subsets: biological; ecological; economic and social, where social includes political and cultural goals. The biological and ecological goals may be more correctly thought of as constraints in achieving desired economic and social benefits but for simplicity and consistency with the terminology most commonly used in fisheries management, we will include them as goals in this Guidebook. Examples of goals under each of these categories include:

- to maintain the target species at or above the levels necessary to ensure their continued productivity (biological);

- to minimise the impacts of fishing on the physical environment and on non-target (bycatch), associated and dependent species (ecological);

- to maximise the net incomes of the participating fishers (economic); and

- to maximise employment opportunities for those dependent on the fishery for their livelihoods (social).

Identifying such goals is important in clarifying how the fish resources are to be used to benefit society, and they should be agreed upon and recorded, both at the policy level and for each fishery. Without such goals, there is no guidance on how the fishery should be operated, which results in a high probability of *ad hoc* decisions and sub-optimal use of the resources (resulting in lost benefits), and increases the probability of serious conflicts as different interest groups jostle for greater shares of the benefits. This is often seen in practice and one of the important causes of failures in fisheries management has been identified as the frequent absence of clear and precise objectives.

While setting goals is an essential first step, the goals stated above have two obvious limitations. Firstly, they have clear conflicts in intention as it is impossible, for example, to minimise impacts of the fishery on the ecosystem and simultaneously to maximise net incomes. Similarly, it is very probable that management strategies that aim to maximise net incomes will not also maximise employment opportunities. Some compromise between these goals has to be achieved before an effective management strategy can be devised. The second limitation of the goals is that they are too vague to be of much benefit to the manager. For example, the impacts of fishing can only be "minimised" by having no fishing at all, which is unlikely to have been the intention of those who stated the goal. Maximising employment opportunities could mean allowing as many fishers as possible to participate, regardless of whether or not they could make a living from the fishery, or it could mean maximising the number which could still earn some acceptable income, or many other such targets. Too much is left to the discretion of the manager with these examples of goals.

It is therefore necessary to refine the goals further and to develop operational objectives for each fishery (Figure 2). Operational objectives are very precise and are formulated in such a way that they should be simultaneously achievable in that fishery. In other words, the trade-offs between the biological, ecological, economic and social goals must have been agreed upon and the conflicts and contradictions resolved. The development of operational objectives is discussed in Chapter 5 but, to illustrate the difference between goals and operational objectives, two examples of objectives taken from Chapter 5 are:

- to maintain the stock at all times above 50% of its mean unexploited level (biological);
- to maintain all non-target, associated and dependent species above 50% of their mean biomass levels in the absence of fishing activities (ecological).

With operational objectives such as these, it is possible for any observer, including the manager, to establish whether or not they are being achieved and hence whether or not the management strategy is appropriate and being successfully implemented. These operational objectives can also easily be used as the foundation for reference points, which are essentially the operational objectives expressed in a way which can be estimated or simulated in a fisheries assessment (Figure 2). Once operational objectives have been agreed upon, a management strategy can be developed, made up of a suite of different management measures, to achieve those objectives.

All of this may sound complex, but in reality is no more than most people do in order to develop a budget for their personal finances. Most of us have realistic but imprecisely expressed hopes and needs for our lifestyle as well as a knowledge of the nature of the resource (in this case our net income). These hopes and needs are the goals of our budget but they will all compete for the same resource, our net income, so there will probably be conflicts which need to be resolved. Therefore we have to modify our goals and express them more precisely: we develop operational objectives in which we specify what we can realistically achieve in terms of food, housing, education etc. Thereafter, we need to decide on our budget strategy: how can we meet those objectives: what type and quantities of food and clothing should we be buying; what type of housing can we consider; can we consider an annual holiday, etc.

Clearly, our operational objectives must be consistent with the yield we can expect from the resource (our income). Normally the process of developing realistic objectives will require trade-offs and most of us find, for example, that we cannot allocate as much for entertainment or holidays as we would like and at the same time make our rental or mortgage payments. Therefore priorities are established and compromises made until eventually we arrive at realistic objectives that balance our desires with our income, and that provide a good guide on how to manage our finances from month to month and in the longer-term. At the end of this, we should have a feasible financial management strategy that, barring totally unexpected events, will have a predictable outcome. If we have done our calculations correctly and responsibly, the strategy should mean we enjoy a reasonable lifestyle without being sued for bankruptcy. This is little different from the basic task, and overall hope, of the fisheries manager!

7. MANAGEMENT PLANS, MEASURES AND STRATEGIES

There is a lot of terminology floating around in fisheries management that, unless clearly understood, can cause further confusion in an already confusing environment. In addition to the words 'goals' and 'operational objectives', the terms management plans, management measures and management strategies will crop up on many occasions in this Guidebook and we need to clarify what we mean by each of them and how they differ.

The Technical Guidelines on Fisheries Management (FAO, 1997) describe a management plan as "a formal or informal arrangement between a fisheries management authority and interested parties which identifies the partners in the fishery and their respective roles, details the agreed objectives for the fishery and specifies the management rules and regulations which apply to it and provides other details about the fishery which are relevant to the task of the management authority." A well formulated management plan should be prepared for every fishery and the Code of Conduct (Paragraph 7.3.3) states that: "Long-term management objectives should be translated into management actions, formulated as a fishery management plan or other management framework." Given the importance of management plans in fisheries, the final chapter in the Guidebook, Chapter 9, is devoted to their development.

As discussed in the previous section, fisheries policy is translated into goals and the goals into objectives that indicate precisely what is expected to be achieved from the fishery. The objectives are achieved through the implementation of a management strategy which will also be a central element of a management plan. The management strategy is the sum of all the management measures selected to achieve the biological, ecological, economic and social objectives of the fishery. It is possible that in a single species fishery a management strategy could consist of a single management measure, such as a specified total allowable catch (TAC), but in practice the great majority of management strategies consist of a number of management measures, encompassing technical, input and output controls and a system of user rights. An effective management strategy, however, should not contain so many management measures that compliance and enforcement become so difficult as to be practically impossible.

A management measure is the smallest unit of the fishery manager's tool kit and consists of any type of control implemented to contribute to achieving the objectives. Management measures are classified as technical measures (Chapters 2 and 3), input (effort) and output (catch) controls (Chapter 4), and any access rights designed around input and output controls (Chapter 6). Technical measures can be sub-divided into regulations on gear-type or gear design (Chapter 2) and closed areas and closed seasons (Chapter 3). A minimum legal mesh size, a seasonal closure of the fishery, a total allowable catch (TAC), a limit on the total number of vessels in a fishery, and a licensing scheme to achieve the limit are all examples of management measures. A substantial part of the Guidebook is intended to assist managers in considering and selecting different management measures for a given fishery.

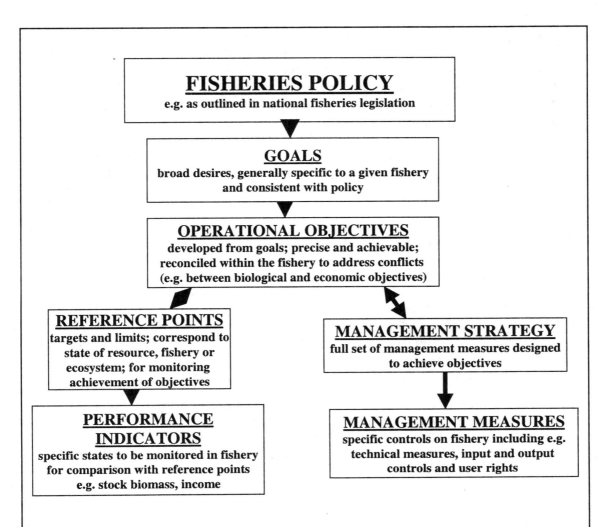

Figure 2 The hierarchical relationships between the different intentions (policy, goal, objectives), standards (reference points and indicators) and actions (management strategy and measures). All of these, and other information as discussed in Chapter 9, would be described in the management plan.

8. PRIMARY CONSIDERATIONS IN FISHERIES MANAGEMENT

If marine living resources were infinite and indestructible, we could leave people to use and abuse them at will. However, this is not the case and we therefore need to manage fisheries to ensure that the resources are utilised in a sustainable and responsible way, and that the potential benefits are not inefficiently dissipated and possibly totally lost. Fisheries production and yield are constrained by a number of factors which can be classified as biological, ecological and environmental, technological, social and cultural, and economic considerations. There are frequently also considerations imposed by other users of the fishing grounds and neighbouring areas. These considerations are discussed in considerable detail in the Technical Guidelines on Fisheries Management (FAO, 1997). Some of the key points are discussed immediately below, but the reader is referred to that publication for more detail. Several of the themes are also dealt with in subsequent chapters of this Guidebook.

8.1 Biological Considerations

As living populations or communities, aquatic living resources are capable of on-going renewal through the processes of growth in size and mass of individuals and additions to the population or community through reproduction (leading to what in fisheries is often called 'recruitment'). In a population at equilibrium, the additive processes of growth and reproduction on average equal the loss process of total mortality. In an unexploited population, total mortality consists only of natural mortality, made up of processes such as predation, disease, and death through drastic changes in the environment. In a fished population, total mortality consists of natural mortality plus fishing mortality, and a primary task of fisheries management is to ensure that fishing mortality does not exceed the amount which the population can withstand, in addition to natural mortality, without undue harm or damage to the sustainability and productivity of the population. This requires not only that the total population is maintained above a certain abundance or biomass, but also that the age structure of the population is maintained in a state in which it is able to maintain the level of reproduction, and hence recruitment, necessary to replenish the losses through mortality. Further, fishing over a long period on selected portions of a stock, for example large individuals or individuals spawning at a specific time or locality within a wider spawning season or range, can reduce the frequency of the particular genetic characteristics giving rise to that feature or behaviour. This has the effect of reducing the overall genetic diversity of the stock. With reduced genetic diversity, the production potential of the population can be adversely affected and it may also become less resilient to environmental variability and change. Fisheries management needs to be aware of this danger and avoid maintaining such selective pressures over a prolonged period.

Achieving an appropriate level and pattern of fishing mortality is hindered substantially by difficulties in estimating population abundance and population dynamics rates and the variability in these rates. The fisheries manager must, however, have sufficient knowledge to make good decisions. The Code of Conduct specifies (Paragraph 7.2.1): "...States... should, *inter alia*, adopt appropriate measures, based on the best scientific evidence available, which are designed to maintain or restore stocks at levels capable of producing maximum sustainable yield, as qualified by relevant environmental and economic factors..." and further (Paragraph 7.5.2, referring to the Precautionary Approach) "In implementing the precautionary approach States should take into account, *inter alia*, uncertainties relating to the size and productivity of the stocks, reference points, stock condition in relation to reference points, ...". These issues are discussed in Chapter 5.

Fishery managers must also respect the stock structure of the resources. Fish populations are frequently made up of a number of different stocks, each of which is genetically largely isolated from the others through behavioural or distributional differences. The different stocks also reflect genetic diversity and if a particular stock is fished to extinction or to very low levels, this genetic diversity may be lost. The stock will not readily be replenished from other stocks, because of the genetic isolation, and therefore the production it was generating will also be lost, leading to a permanent or at least long-term loss of benefits. Fisheries management should therefore attempt to address each stock separately and to ensure sustainable use of each stock and not just of the population as a whole. In this regard, the Code of Conduct states (Paragraph 7.3.1): "To be effective, fisheries management should be concerned with the whole stock unit over its entire area of distribution and take into account previously agreed management measures established and applied in the same region, all removals and the biological unity and other biological characteristics of the stock."

8.2 Ecological and Environmental Considerations

The abundance and dynamics of a population place an important constraint on fisheries but aquatic populations do not live in isolation. They exist as components of a frequently complex ecosystem, consisting of biological components which may feed on, be fed on by, or compete with a given stock or population. Even those populations which are not directly linked through the food web may indirectly affect each other through their direct interactions with predators, prey or competitors of the other. The physical component of the ecosystem, the water itself, the substrate, inflows of freshwater or nutrients and other non-biological processes may also be very important. Different substrates may be essential for the production of food organisms, for shelter, or as spawning or nursery grounds.

The environment of fish is very rarely static and conditions, particularly of the aquatic environment, can vary substantially over time, from hourly variability, such as the tides, to seasonal variability in, for example, water temperature and currents, to decadal variability as in the occurrence of El Niño events and regime shifts. These changes frequently affect the population dynamics of fish populations, resulting in variability in growth rates, recruitment, natural mortality rates or any combination of these. Such variability can also affect the availability of fish resources to fishing gear, not only affecting the success of the fishing industry, but also the way in which the fishery scientist must interpret catch and catch rate information from the fishery.

Changes in any of the biological, chemical, geological or physical components of the ecosystem can have impacts on the resource population and community. Some of these changes may be beyond human control, such as upwelling processes enriching some coastal ecosystems or large scale temperature anomalies, but they still need to be considered in the management of the resource. Others, such as the destruction of coastal habitats for development, or the direct impact of fishing on the substrate or on other species impacting the resources, are due to human action. In these cases, fisheries management should both take into account their impacts on the resource and, in consultation with other relevant agencies and parties, take steps to minimize their impacts on the fishery ecosystem.

The manager also needs to consider the impact of the fishery on the ecosystem as a whole (Code of Conduct, Paragraph 7.2.2 g and 7.6.9). There are four types of impact of fisheries on the ecosystem: direct impact on the target species; direct impacts on the bycatch species (including discards and by-mortality – Chapter 2); indirect impacts on other organisms transmitted through the food chain (i.e. by changing the abundance of predators, prey or competitors of a population); and direct impact of fishing on the physical or chemical environment. The manager needs to be aware of these potential impacts and to use management measures that minimise negative impacts.

The potential to address the ecosystem considerations will vary depending on whether they are caused by or independent of human action, but in both cases the constraints imposed on the resources and the fishery by biological and non-biological ecosystem factors need to be recognised. At the most fundamental level, these factors in combination with the biology of the species determine the maximum abundance, or carrying capacity, and productivity of the resources. Changes in the ecosystem can affect both and, where they are occurring, need to be considered by the fisheries manager.

Again, these aspects are dealt with by the Code of Conduct. Amongst several references, Paragraph 7.2.3 specifies "States should assess the impacts of environmental factors on target stocks and species belonging to the same ecosystem or associated with or dependent upon the target stocks, and assess the relationship among the populations in the ecosystem." and Paragraph 7.6.9 affirms "States should take appropriate measures to minimize waste, discards, catch by lost or abandoned gear, catch of non-target species, both fish and non-fish species, and negative impacts on associated or dependent species, in particular endangered species.

8.3 Technological Considerations

The fishery manager has very little, if any, ability to influence directly the dynamics of the fish populations or communities which support a fishery. In some case, particularly inland waters, there may be opportunities and a desire to undertake stock and habitat enhancement and in some coastal fisheries, habitat destruction may have had an impact on fish production. In the latter case, restoration or stabilization may well be an issue the fisheries manager needs to consider (Code of Conduct, Paragraph 7.2.2 f) and Article 10). However, in most fisheries, the only mechanism the fishery manager has to ensure sustainable utilization of the resources is by regulating the quantity of fish caught, when and where they are caught and the size at which they are caught. This can be done through directly regulating the catch taken, by regulating the amount of effort allowed in the fishery, by specifying closed seasons and closed areas and by regulating the type of gear and fishing methods used. However, there are constraints on how precise the manager can be in setting such regulations. Catch controls are often difficult to monitor and therefore to implement. It is difficult to estimate fishing effort precisely, and normally improving technology and developing skills result in on-going increases in the efficiency of fishing operations, leading to continuing increases in effective effort, unless steps are actively taken to counter these improvements or their consequences. Fishing gear is rarely strongly selective and bycatch of non-target species or unwanted sizes of target species is frequently a problem. The uncertainties in fisheries management are not just at the level of predicting the status and dynamics of the resources, and uncertainties in the real consequences of implementing fishery measures is also a significant problem to the manager. The properties of these measures and when and how to use them is dealt with in considerable detail in subsequent chapters, especially Chapters 2 to 4.

A fundamental problem in many fisheries is the existence of too much effort. The presence of excess effort will frequently result in on-going pressure on the fisheries manager to exceed the sustainable fishing mortality on a resource. The social and political pressure to provide employment and opportunities for all those with a stake in the fishery is often hard to resist and readily leads to over-exploitation. The Code of Conduct requires that States take measures to prevent or eliminate excess fishing capacity (Code of Conduct, Paragraph 7.1.8) and such is the global level of concern that the FAO members have agreed on an International Plan of Action (IPOA) for the Management of Fishing Capacity[2].

8.4 Social and Cultural Considerations

Human populations and societies are as dynamic as other biological populations, and social changes take place continuously and on different scales, affected by changes in weather, employment, political circumstances, supply of and demand for fisheries products and other factors. Such changes can affect the appropriateness and effectiveness of management strategies, and therefore need to be considered and accommodated by them. However, again as with biological and technological factors, it can be difficult to identify and quantify the key social and cultural factors influencing fisheries management, generating additional uncertainties for the manager.

A major social constraint in fisheries management is that human societies and behaviour are not easily transformed and fishing families and communities may not be willing to move into other occupations, or away from their normal homes when there is surplus capacity in a fishery, even when their quality of life may be suffering as a result of depleted fish resources. The problem is much worse when there are no other opportunities outside of fisheries in which they could earn a basic living. Under such circumstances, the political decision to reduce capacity in the fishery is an extremely unattractive option, as the short-term costs of excluding dependent people from the fishery will be much more visible and hence unpopular than a "hands-off" approach which allows the resource and fishery to dwindle in magnitude and quality under sustained excess fishing mortality. Nevertheless, the ecological, economic and social consequences of the latter choice are far more serious in the longer term. This reluctance or inability to take decisions with

---

[2] The details of the IPOA can be found at http://www.fao.org/fi/ipa/capace.asp

serious, immediate social consequences for some has been one of the constraints most responsible for over-fishing around the world.

A key requirement for ensuring that social and cultural considerations are properly considered is to involve the interested parties in fisheries management, keeping them well-informed on the management aspects of the fishery and providing them with the opportunity to express their needs and concerns. This is discussed in Chapter 7 of the Guidebook. The Code of Conduct (Paragraph 7.2.2) suggests that "the interests of fishers, including those engaged in subsistence, small-scale and artisanal fisheries are taken into account;" and (Paragraph 7.1.2) "Within areas under national jurisdiction, States should seek to identify relevant domestic parties having a legitimate interest in the use and management of fisheries resources and establish arrangements for consulting them to gain their collaboration in achieving responsible fisheries".

The relative balance between social and economic considerations in a fishery will depend on the priority given by the appropriate authority to social objectives and economic objectives. Social and economic objectives can conflict: for example it is unlikely that maximising economic efficiency and maximising employment could be simultaneously pursued within a given fishery, and attempting to do so will result in conflict. A common example of such conflicts is that between a commercial fleet pursuing essentially economic objectives and an artisanal fleet fulfilling primarily social objectives, with both having an impact on the same stock, and possibly also interfering with each other's fishing operations. It is important for the management authority to have identified such potential conflicts and to have resolved them, identifying and specifying compromise objectives that achieve general support.

8.5 Economic Considerations

In a fishery for which sustainable economic efficiency had been specified as the sole benefit to be extracted and in which optimum circumstances prevailed, market forces could be anticipated to lead to the desired objective of economic efficiency. However, in reality such optimum conditions are rarely if ever found and uncertainties and externalities distort the natural selection of market forces. Uncertainties include unpredictable variability in resources and other sources of imperfect information, and externalities can include the impacts of other fisheries on the target resources (e.g. taking them as bycatch), subsidies, trade regulations, fiscal regulations and variability in markets and demand. All of these introduce complexity and additional uncertainty into a fishery and, without proper management, will lead to sub-optimum economic performance. It is important for the management authority to consider the broad economic context of a fishery, including relevant macroeconomic factors. As with social considerations, this requires close consultation with the legitimate users who will be the ones most affected by and sensitive to these issues.

At one extreme, although still very common in fisheries especially in many developing countries, are the problems of open access fisheries, in which anyone is allowed entry into a fishery. Under these circumstances, people will continue to enter the fishery until the benefits from fishing are so low as to be unattractive to prospective new entrants (Section 2, Chapter 6). How low this is will depend largely on the availability of other options and in many countries, especially developing countries, such alternatives may be extremely scarce. Even where there are reasonable alternatives, the inevitable result of open access fisheries is dissipation of rent leading to very poor economic efficiency and, unless strong and effective management measures are in place and enforced, to over-exploitation of resources. Such circumstances prevail in many fisheries around the world.

Recognizing this most elementary lesson in fisheries management, the Code of Conduct calls for the adoption of "measures to ensure that no vessel (by which should also be understood no shore-based fisher) be allowed to fish unless so authorised.." (Paragraph 7.6.2), that "States should ensure that the level of fishing is commensurate with the state of fisheries resources."

(Paragraph 7.6.1), and going further than that, that "Where excess fishing <u>capacity</u> exists, mechanisms should be established to reduce capacity to levels commensurate with the sustainable use of fisheries resources so as to ensure that fisheries operate under economic conditions that promote responsible fisheries." (Paragraph 7.6.3), where underlining and bracketed comments on Code text are additions by the Chapter author. Taken together, these three paragraphs specify that responsible fisheries require limited and authorised access by fishers, where actual and potential effort is appropriate to the productivity of the resource or resources being exploited.

8.6 Considerations Imposed by Other Parties

Some offshore fisheries operate in effective isolation from any other users and the regional fisheries organisations charged with their management may be able to manage the fisheries without needing to consider conflicts with or interference from non-fishery users. However, the bulk of global fishery landings come from coastal waters and for many if not most of the fisheries producing these landings, other users are an important consideration and frequently a constraint. Other users of the fishing grounds can include, for example, tourism, conservation, oil and gas extraction, offshore mining and shipping, while use of the intertidal and coastal area can include tourism again, aquaculture and mariculture, coastal zone development for housing, business or industry, and agriculture. All of these can impose significant constraints on fishing activities and may be impacted by fishing activities. The manager therefore needs to be aware of such activities and of real or potential impacts in both directions. When developing management strategies and formulating management measures, potential conflicts with other users need to be identified and addressed, and the potential impacts of other users on the efficacy of the management strategy and measures need to be considered. The strategy must be adapted so as to account for and be robust to these impacts.

An unavoidable implication of overlapping interests is that the fishery manager, through the management authority, must ensure that suitable structures and mechanisms are put into place for effective communication and decision-making with representatives of the other users. In addition to reference in Paragraphs 6.8 and 6.9, this is dealt with mainly in Article 10 of the Code of Conduct: Integration of Fisheries into Coastal Area Management, which includes the requirement that (Paragraph 10.4.1): "States should establish mechanisms for cooperation and coordination among national authorities involved in planning, development, conservation and management of coastal areas."

9. IS ALL THIS LEGAL?

It should not need to be stated that it is essential for the fisheries manager to be thoroughly conversant with the laws and regulations which control the fisheries within his or her jurisdiction. These laws and regulations constitute the legal regime under which the fishery should be operated and managed, and include the national legislation and any relevant international legal instruments (Table 2). The term legislation is used here to include all types of national laws, local laws, regulations and customs.

9.1 National Legislation

The scope of the national legislation varies substantially between countries, depending on, for example, whether a particular country has a common law system, a civil law system or any other system. However, typically the primary legislation is broad, prescribing the principles and policy relating to fisheries and is usually approved by the Legislature of that country, which may be the national Congress or Parliament. It may also specify details on the implementation of aspects of the policy considered to be particularly important or sensitive and should include reference to establishing fishery management plans and the procedures for the planning process. This primary legislation would usually be described in a Fisheries Act or equivalent legislation.

As passage through the legislature is usually a slow process, such primary legislation should normally not need to be changed frequently. Therefore, for example, control measures such as the amount of effort allowed in a given fishery, or the annual TAC should not be included in the primary legislation.

The primary legislation would typically provide the legal basis for development of detailed procedures and regulations for its implementation by a designated law-making authority. The delegated powers should define and empower the designated institutional components responsible for fisheries management, including specification of who is responsible for administration and control of fisheries management. The second-tier laws, or so-called subsidiary legislation, produced by the delegated regulatory authority are often referred to as regulations, orders, proclamations etc. They would include specifying the control measures which require frequent, typically annual, revision such as licences, gear restrictions, closed areas and seasons and input and output controls (Chapters 2 to 4).

9.2 International legislation and instruments

The modern fisheries manager is required to be familiar not only with the national legislation governing fisheries, but also with the bewildering diversity of international legislation and voluntary instruments dealing directly with or impinging on fisheries. There has been a proliferation of such instruments in recent decades and a few of the more important examples and types are listed in Table 2.

Chief amongst the international instruments is the United Nations Convention on the Law of the Sea of 10 December 1982 (LOS Convention), which entered into force in 1994 (Table 2). This convention sets the legal context for all subsequent international arrangements and agreements relating to the use of the oceans and seas. Arising directly from the LOS Convention and designed to strengthen its provisions relating to high seas fisheries and transboundary stocks, are the UN Fish Stocks Agreement and the FAO Compliance Agreement.

There is also a host of other global agreements, both binding and voluntary. To date the Convention for International Trade in Endangered Species of Fauna and Flora (CITES) has had little impact on marine fisheries management, but concern about some marine species subjected to international trade is growing. Given this growing attention, there is a high likelihood that more species of fisheries interest will be listed through CITES in the future. For example, sturgeon species (Acipenseriformes spp.) are currently listed on Appendix II, under which international trade is carefully monitored and controlled, and the basking shark was placed on Appendix III of CITES by the United Kingdom in 2001. Some other global instruments of more immediate relevance are also shown in Table II, including the Convention on Biological Diversity.

Most countries involved in fisheries are or will become members of one or more regional bodies involved in utilization, management and conservation of marine living resources. These include bodies such as the various tuna commissions (e.g. the International Convention on the Conservation of Atlantic Tuna (ICCAT) and the Convention on Indian Ocean Tuna (IOTC)), the Convention on the Conservation of Antarctic Living Marine Resources (CCAMLR), various FAO regional fishery bodies such as the Fishery Committee for the Eastern Central Atlantic (CECAF) and the Asia-Pacific Fishery Commission (APFIC), and many others. The manager must be aware of those in which his or her country is involved, and the implications and obligations of membership.

10. RECOMMENDED READING

Berkes, F, R. Mahon, P. McConney, R. Pollnac & R. Pomeroy. 2001. *Managing Small-scale Fisheries. Alternative Directions and Methods*. IDRC, Canada. 320 pp.

Caddy, J.F. and Griffiths, R.C. 1995. Living marine resources and their sustainable development. Some environmental and institutional perspectives. *FAO Fisheries Technical Paper*, **353**. 167 pp.

Charles, A. T. 2001. *Sustainable Fishery Systems*. Blackwell Science, London. 384 pp.

Cochrane, K.L. 2000. Reconciling sustainability, economic efficiency and equity in fisheries: the one that got away? *Fish and Fisheries*, **1**: 3-21.

FAO. 1995. Code of Conduct for Responsible Fisheries. FAO, Rome. 41pp.

FAO. 1997. FAO Technical Guidelines for Responsible Fisheries No. **4**: Fisheries Management. FAO, Rome. 82pp.

FAO. 2000. *The State of World Fisheries and Aquaculture*. 2000. FAO, Rome.

United Nations. 1998. *International Fisheries Instruments with Index*. Division for Ocean Affairs and the Law of the Sea, Office of Legal Affairs. United Nations, New York. 110 pp.

Table 2. Some key legislation and agreements which make up the legal regime of fisheries management.

Law or Agreement	Comments
a) Legislation directly pertaining to fisheries	
- The primary national legislation relating to fisheries (e.g. a National Fisheries Act)	
- The secondary legislation pertaining to specific fisheries and control measures, including regulations and, where appropriate, traditional customs and practices.	
- United Nations Convention on the Law of the Sea of 10 December 1982	Entered into force on 16 November 1994. Provides a comprehensive regime of law and order in the world's oceans and seas establishing rules governing all uses of the oceans and their resources. It enshrines the notion that all problems of ocean space are closely interrelated and need to be addressed as a whole.
- Agreement for the Implementation of the Provisions of the United Nations Convention on the Law of the Sea of 10 December 1982 Relating to the Conservation and Management of Straddling Fish Stocks and Highly Migratory Fish Stocks (1995 UN Fish Stocks Agreement)	The 30th ratification or accession necessary for it to enter into force was received on 11 November 2001. The Agreement elaborates on LOS Convention principle that States should cooperate to ensure conservation and promote the objective of the optimum utilization of fisheries resources both within and beyond the exclusive economic zone.
- Agreement to Promote Compliance with International Conservation and Management Measures by Fishing Vessels on the High Seas (FAO Compliance Agreement)	As of November 2001 not yet in force: 22 of the 25 acceptances required had been received. Addresses the problems associated with reflagging of fishing vessels as a means of avoiding compliance with applicable conservation and management rules for fishing activities on the high seas.
Convention on Biological Diversity	A binding agreement: countries that are parties to the Convention are obliged to implement its provisions. Has three main goals: conservation of biodiversity; sustainable use of the components of biodiversity; and sharing the benefits arising from utilization of genetic resources in a fair and equitable way.
- Any obligations imposed by international organisations of which the State is a signatory e.g. CITES, International Whaling Commission (IWC), tuna commissions, etc.	

Law or Agreement	Comments
- Any relevant, binding bilateral or multilateral fisheries agreements	
b) Voluntary Agreements pertaining to fisheries	
- FAO Code of Conduct for Responsible Fisheries	Adopted by the Twenty-eighth Session of the FAO Conference on 31 October 1995. Sets out principles and international standards of behaviour for responsible practices with a view to ensuring the effective conservation, management and development of living aquatic resources, with due respect for the ecosystem and biodiversity.
- FAO International Plans of Action (IPOAs) for: • Reducing Incidental Catch of Seabirds in Longline Fisheries; • the Conservation and Management of Sharks; • the Management of Fishing Capacity; • to Prevent, Deter and Eliminate IUU Fishing.	The four IPOAs are voluntary instruments elaborated within the framework of the Code of Conduct for Responsible Fisheries. They apply to all States and entities and to all fishers.
- Agenda 21 of the United Nations Conference on Environment and Development	Especially Chapter 17: Protection of the Oceans, All Kinds of Seas, Including Enclosed and Semi-Enclosed Seas, and Coastal Areas and the Protection, Rational Use and Development of Their Living Resources.

CHAPTER 2

THE USE OF TECHNICAL MEASURES IN RESPONSIBLE FISHERIES: REGULATION OF FISHING GEAR

by

Åsmund BJORDAL

Institute of Marine Research, Bergen, Norway

1.	INTRODUCTION	21
2.	FISHING GEARS	22
2.1	The ideal fishing gear	22
2.2	Classification of fishing gears	22
3.	PASSIVE FISHING GEARS	22
3.1	Nets	23
3.2	Hook and line fishing	25
3.3	Pots and traps	28
4.	ACTIVE FISHING GEARS	30
4.1	Spears and harpoons	30
4.2	Trawls and dredges	31
4.3	Seine nets	33
4.4	Beach seines	34
4.5	Purse seines	35
4.6	Other fishing gears and devices	36
5.	GEAR SELECTIVITY AND ECOSYSTEM EFFECTS OF FISHING	36
5.1	Selectivity properties and ecosystem effects of different fishing methods	39
6.	MANAGEMENT CONSIDERATIONS: SELECTIVITY AND OTHER ECOSYSTEM EFFECTS OF FISHING	45
7.	RECOMMENDED READING	46

1. INTRODUCTION

The need for fisheries management arises as the surplus production from fish stocks is overtaken by the catching capacity of fishing fleets. Catching capacity is the product of the fishing effort and the combined efficiency of the fishing gear and the fishing vessel (e.g. loading capacity, engine power, range capacity, fish finding- and navigational equipment) as well as the skills of the crew.

Fisheries management includes different management measures. Among these are technical regulations on fishing gears in order to obtain the overall goal of high sustainable yield in the fisheries. These are regulations e.g. on mesh size to improve the selective properties of a fishing gear so that bycatches of juvenile fish are reduced - in order to safeguard recruitment to the larger size groups of a fish stock including the spawning stock.

In recent years there has been a growing focus on "ecosystem effects of fisheries", addressing the impact of fishing operations not only on the target species, but also on bycatch of or other effects on non-commercial species or habitats. Energy efficiency, reduced pollution and improved quality of the catch are also important aspects related to fishing gears and fishing operations (Code of Conduct for Responsible Fisheries, Article 7.2.2). From a situation where the development of fishing gears and methods only focused on the highest possible catching efficiency for the target species, now fisheries research, fisheries management and the fishing industry are challenged to develop gear, methods and regulations that meet the different considerations mentioned above. This is part of an emerging ecosystem approach to fisheries management.

2. FISHING GEARS

2.1 The ideal fishing gear

Some criteria for the ideal fishing gear could be:

- highly selective for the target species and sizes, with negligible direct or indirect impact on non- target species, sizes and habitats (Code of Conduct, Paragraphs 7.2.2, 8.4.7, 8.5.1 – 8.5.4) ;
- effective, giving high catches of target species at lowest possible cost;
- quality orientated, producing catches of high quality (Code of Conduct, Paragraph 8.4.4).

According to these and additional criteria that could be added to the list, it can easily be stated that the ideal fishing gear does not exist, as no fishing gear fulfils the complete list of desired criteria and properties. However, in the process of moving towards sustainable fisheries management, different fishing gears with their specific properties and potential for improvement are an important compartment in the "fisheries manager's toolbox". A basic understanding of the properties, function and operation of the major fishing gears and methods is therefore fundamental for decision-making in fisheries management, particularly when it comes to technical measures in fisheries regulations.

2.2 Classification of fishing gears

Fishing gears are commonly classified in two main categories: passive and active. This classification is based on the relative behaviour of the target species and the fishing gear. With passive gears, the capture of fish is generally based on movement of the target species towards the gear (e.g. traps), while with active gears capture is generally based on an aimed chase of the target species (e.g. trawls, dredges). A parallel on land would be the difference between the trapping of and hunting for animals.

In the following sections a short description of the major gear types is given, including their catching principle, construction, operation and common target species. Gear selectivity and properties related to ecosystem effects of fishing will be treated in Section 5.

3. PASSIVE FISHING GEARS

Passive gears are in general the most ancient type of fishing gears. These gears are most suitable for small scale-fishing and are, therefore, often the gear types used in artisanal fisheries. Some passive fishing gears are often referred to as "stationary" fishing gears. Stationary gears are those anchored to the seabed and they constitute a large group of the passive gears. However, some moving gears such as drift nets may also be classified as passive gears, as fish capture by these gears also depend on movement of the target species towards the gear.

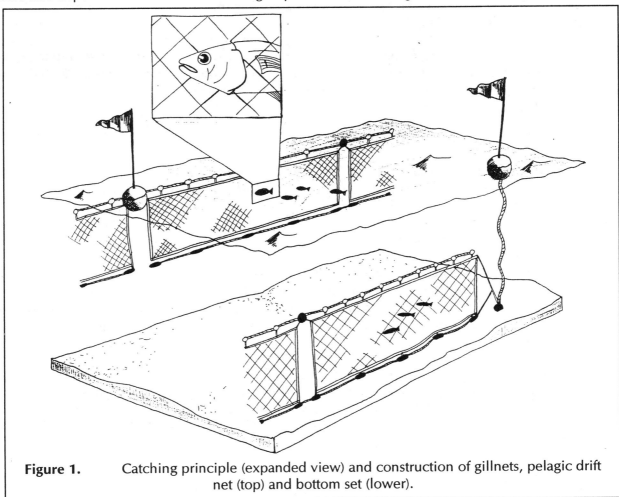

Figure 1. Catching principle (expanded view) and construction of gillnets, pelagic drift net (top) and bottom set (lower).

3.1 Nets

(a) Gillnets

The catching principle and construction of gillnets are shown in Figure 1.

Catching principle

The gillnet is named after its catching principle, as fish are usually caught by "gilling" – i.e. the fish is caught in one of the meshes of the gillnet, normally by the gill region (between the head and the body). Thus, fish capture by gillnets is based on fish encountering the gear during feeding or migratory movements. As fish may avoid the gillnet if they notice the gear, catches are normally best at low light levels or in areas with turbid water.

Construction

A gillnet consists basically of a "wall" or panel (e.g. 5 by 30 m) of meshes made from fine thread. The mesh panel is mounted with reinforcing ropes on all sides. To obtain a vertical position of the net in the sea, floats and weights are fastened at regular intervals to the top rope (float line, cork line) and bottom rope (sinker line, lead line), respectively. The size of the meshes and hanging ratio (number of meshes per length of gillnet) are chosen to fit the desired target species and size.

Mesh size is commonly given as the length (in mm) either of a whole stretched mesh or the half-length (also called bar-length).

Today, gillnets are almost exclusively made from synthetic fibres, normally nylon (polyamide) – either as multifilament thread or monofilament (gut). The latter is increasingly being used because of its low visibility and correspondingly higher catch efficiency. Multi-monofilament is also becoming more common.

Operation

Gillnets are most commonly operated as a stationary gear anchored to the bottom at either end, but may also be so-called drift-nets which float freely in the water. Stationary nets may be set on the seabed, at different depths in the water column or with the float line at the surface. Similarly, drift-nets may be operated with the float line at the surface or suspended from surface floats and corresponding float lines to the desired fishing depth in midwater.

Gillnets may be operated from vessels ranging from the smallest non-mechanised fishing boats to big, well equipped vessels capable of large-scale deep sea fishing. The gear used in small- and large-scale fishing is basically the same: the unit gillnet. However, with increased vessel size a larger number of net units can be carried and operated per day. Single gillnets are then linked into long fleets of up to several hundred nets.

Gillnets can also be operated from shallow to large depths and can be used for fishing on rough bottom and at wrecks. One specific problem with gillnets is so-called "ghost fishing". This refers to gillnets that are lost (most commonly after being stuck on a rough bottom) and continue to catch and kill fish over long periods of time. The Code of Conduct (Paragraph 7.2.2) requires that the incidence of ghost fishing should be minimised.

Target species

Gillnets are used to catch a large variety of fish species. In general, bottom gillnets are used for catching demersal species like cod, flatfish, croakers and snapper, while pelagic gillnets are used for species like tuna, mackerel, salmon, squid and herring.

(b) Trammel nets

The catching principle and construction of trammel nets are shown in Figure 2.

Catching principle

In trammel nets, fish are caught by entanglement, which is facilitated by its special construction of three panels of nets attached on the same rope with a high degree of slackness.

Construction

At first glance, a trammel net may look like a gillnet. However, while the gillnet has a single panel of meshes, the trammel net has three – one middle panel of small meshes and two side panels of larger meshes. When a fish comes in contact with the net, it will press the small mesh net through an adjacent larger mesh so that it is caught by entangling or "pouching".

Operation

Trammel nets are usually set and operated like bottom set gillnets, mainly in small-scale, near-shore fisheries.

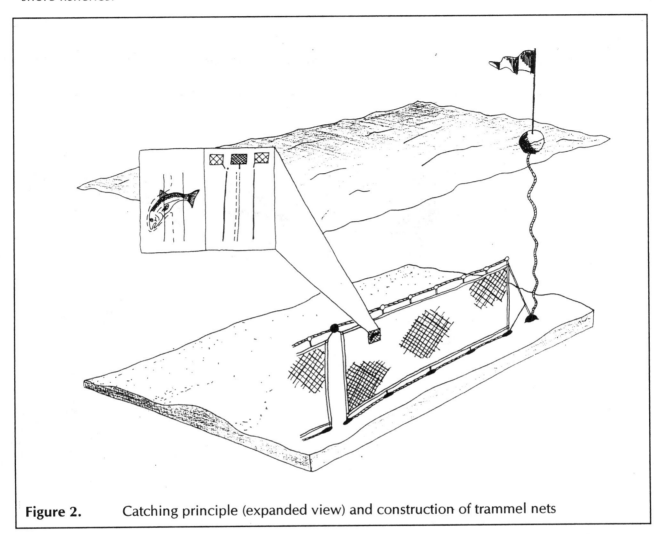

Figure 2. Catching principle (expanded view) and construction of trammel nets

Target species

Trammel nets are used for catching a large variety of demersal fish.

3.2 Hook and line fishing

Different fishing methods are based on the use of fish hooks; longlining, trolling and various forms of handlining such as jigging. The general catching principle of hook fishing is to attract the fish to the hook and entice the fish to bite and/or swallow the hook so that the fish becomes hooked and retained.

(a) Handlining and trolling

The catching principle and construction of handlining are shown in Figure 3.

Catching principle

The fish is attracted to the hook by visual stimuli, either natural bait or more commonly in the form of artificial imitations of prey organisms like lures, jigs, rubber worms etc.

Construction

The gear is simple: a nylon monofilament is commonly used as line with one to several hooks at the end with bait or lures.

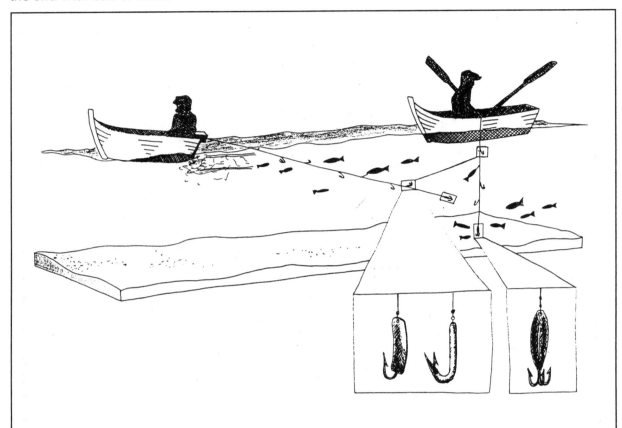

Figure 3. Catching principle and construction of trolling (left) and jigging (right). Expanded view: examples of lures and jig.

Operation

In handlining the fishing line is vertical and is operated from a drifting or anchored vessel. Handlining is also conducted from the shore, with and without the use of a pole. From using only a single line, the operation can be scaled up by using several lines on larger vessels. In recent years jigging has become mechanised and automated by the development of jigging machines.

Hook and line can also be used in trolling where the fishing line is towed behind the moving vessel. Semi-automation has also been developed in trolling where power reels are often used for hauling the lines. Trolling is considered to be a separate type of fishing gear from handlining in the International Standard Statistical Classification of Fishing Gear (Nédélec and Prado, 1990).

Target species

Typical target species with handlining are demersal fishes like cod and snapper as well as squid. Trolling is mainly directed towards pelagic species like mackerel, tuna and salmon.

(b)   Longlining

The catching principle and construction of longlines are shown in Figure 4.

Catching principle

Longlining is based on attracting fish by bait attached to the hook. While handlining and trolling generally exploit the visual sense of the fish to attract it to the hook by artificial lures, longlining exploits the chemical sense of the fish. Odour released from the bait triggers the fish to swim towards and ingest the baited hook with a high probability of being caught.

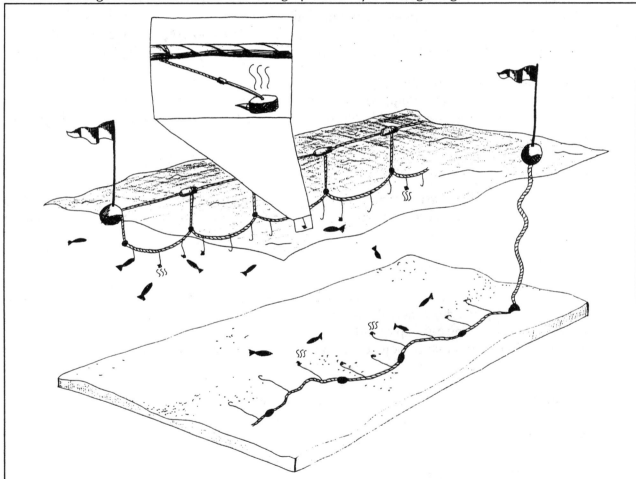

Figure 4.   Catching principle and construction of longlines. Pelagic/drift (top) and bottom set (lower). Expanded view: baited hook connected with gangion (snood, branchline) to mainline.

Construction

As the name of the gear indicates, this is a long line (mainline) with baited hooks attached at intervals – connected to the mainline with relatively shorter and thinner leader lines (snoods, gangions). Depending on the type of fishery, there are great variations in the gear parameters, such as thickness and material of main and leader lines, the spacing between hooks, as well as hook and bait types.

Today, main and leader lines are almost exclusively made from synthetic materials like polyamide (nylon) or polyester. Multifilament (rope) is generally used for main and leader lines with demersal longlines (set on the bottom), while monofilament (gut) is commonly used in pelagic longlining. Hook type (size and shape) varies greatly with target species. Naturally, larger hooks and correspondingly stronger main and leader lines are used for larger fish. There is also a great variation in baits used in different longline fisheries, but the major types of bait are either different pelagic fish (e.g. herring, mackerel, sardine, saury) or different species of squid.

Operation

The longline fishing cycle includes the following main operations: baiting (threading a piece of bait on each hook), setting, fishing ("soaking" the line for some hours), retrieval, removal of fish and old bait, gear maintenance, baiting, etc.

As with gillnets, the gear is basically the same in small and large-scale operations with the length of the line and number of hooks increasing with vessel size. Small, open vessels normally fish a few hundred hooks, while the largest longline vessels (LOA 50-60 m) may operate 50-60 km of longline and as many as 40-50 000 hooks per day.

With increased vessel size there is normally an increased degree of mechanised gear handling. Most longline vessels are equipped with power haulers. In so called auto-lining the laborious baiting process is also mechanised with machines that can bait up to four hooks per second as the line is set into the sea.

Target species

Pelagic (drifting) longlines are typically used for catching species like tuna, swordfish and salmon, while bottom set longlines are used for demersal species like snapper, cod, haddock, halibut, ling, tusk, hake and toothfish.

3.3 Pots and traps

Pots are considered within the International Standard Statistical Classification of Fishing Gear to be a type of trap (Nédélec and Prado 1990) but are described separately here because of the differences in catching principle and construction between pots and other forms of trap. The general catching principle of pots (creels) and traps is to entice or lead the target species into a box or compartment from which it is difficult or impossible to escape.

The catching principle and construction of pots and traps are shown in Figure 5.

(a)   Pots

Catching principle

As with longlining, pot fishing is normally based on attracting target organisms by bait (chemical stimuli). When attracted to the pot, the target organism must enter the pot to gain access to the bait. This can be done through one or several entrances (funnels) of the pot.

Construction

Typical pot shapes are box, cone, cylinder, sphere or bottle. The size of pots may vary from small crayfish pots (conical: 0.3 m diameter and 0.2 m height) to large king crab pots (box shaped: 2x2x1 m). The pot entrances are usually funnel- or wedge-shaped so that the target organism is led into the pot fairly easily, but with low probability of escaping. Pots may be constructed from various materials like wood, palm leaves, metal frames lined with webbing, wire mesh or plastic materials.

Operation

Pots are normally set on the bottom, either as single pots with a buoy line to the surface or in strings of several pots connected to a main line at certain intervals. Pot gear is usually soaked overnight, but longer soak times may be used in certain fisheries. The operation cycle is similar to that of longlining, with baiting, setting, fishing and retrieval. The bait is either freely suspended in the middle of the pot, or put in perforated bait containers to prevent it from being eaten by scavengers. As in longlining, different pelagic species like sardines, herring and mackerel are typically used for pot bait, but most kinds of fish and mussels etc. may be used.

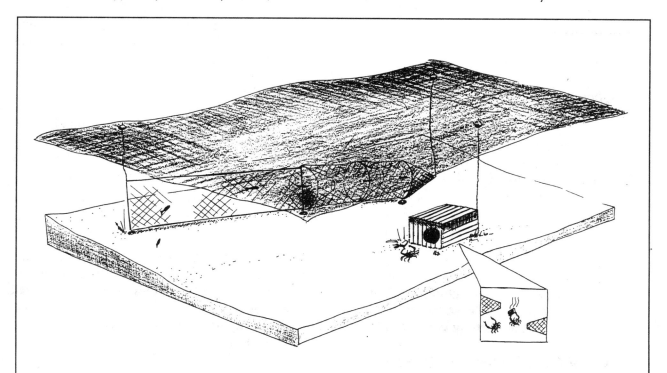

Figure 5. Catching principle and construction of pots (lower) and traps (fyke net - top). Expanded view: schematic illustration of entrances (funnels) and bait.

Target species

Pots are most widely used to catch different crustaceans, like crabs, lobsters and shrimps. Pots are also used for catching different species of finfish like sablefish, tusk and cod in temperate waters and reef fish such as groupers in tropical waters. Other species that are caught with pots are whelks and octopus.

(b) Traps

Catching principle

Traps are normally not baited, but catch fish and other organisms by leading them into the trap, eventually to the fish compartment, that is designed for holding the fish entrapped with low possibility of escaping.

Construction

Compared with pots, traps are usually larger and often more permanent constructions. Tidal traps are based on walls or fences forming V-shaped constructions that entrap fish that have come in with the tide as the tide goes out. Typical salmon and cod traps are cage-like constructions made from webbing with long leader nets to guide the migrating fish into the trap. Tuna may also be caught with traps of this type. The fyke net (Figure 5) is a smaller form of trap, with a leader net connected to the trap or "fyke" which usually consists of three compartments with funnels leading from the outer to the middle and finally to the inner compartment or "fish bag".

In tidal traps the fish are confined sideways by the leader walls, above by the sea surface and below by the sea bed. In floating fish traps like those used for cod and salmon, the fish are confined by side and bottom net panels while the sea surface serves as the top confinement. Fyke nets are set under the surface and thus the fyke (trap) has to be completely lined with webbing to confine the fish.

Operation

Tidal traps are usually permanent constructions where the fish compartment is emptied at low tide. Cod and salmon type traps are usually set out for the season and operated for one to a few months, emptying the fish compartment on a daily basis. Fyke nets are operated like pots, set one by one in the littoral zone, from one to ten meters depth, and are usually moved to another spot after retrieval, usually on a daily basis.

Target species

A variety of target species are caught by tidal traps, both finfish and crustaceans, e.g. shrimps, naturally dominated by species living in the tidal zone. As mentioned above, traps are traditionally used for catching cod and salmon (N. Atlantic), tuna (Mediterranean), small pelagic species in Far East Asia, some species of weakfish (members of the Sciaenidae) and others. Fyke nets are used for catching various species, but are particularly used for eel and cod.

4. ACTIVE FISHING GEARS

Fish capture by active gears is based on the aimed chase of the target species and combined with different ways of catching it.

4.1 Spears and harpoons

This is one of the most ancient ways of active fish capture.

The catching principle and construction of spears and harpoons are shown in Figure 6.

Catching principle

Capture with spears and harpoons depends on visual observation of the target species, which is then impaled by the spear or harpoon from a relatively short distance.

Construction

Basically, the spear or harpoon is designed for easy penetration of the target organism, but the spear head is equipped with barbs or flukes that hold the prey when it is hit. Usually the spear or harpoon is connected to the fisher and boat by a line, so that it can be retrieved, with or without catch.

Operation

Spears and harpoons are most often operated from a vessel, but can also be used from land.

Target species

Common target species with this fishing method are flatfish, swordfish, tunas and whales.

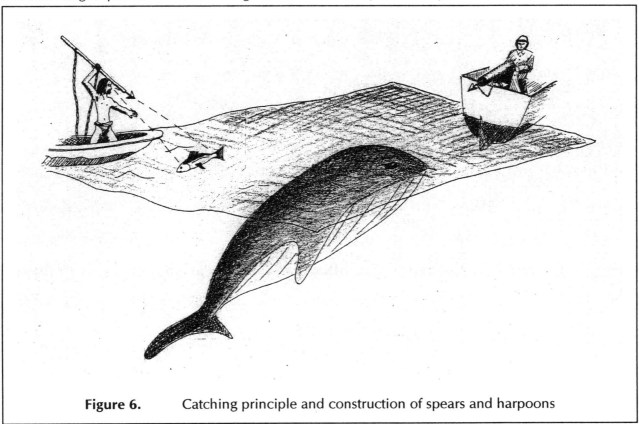

Figure 6. Catching principle and construction of spears and harpoons

4.2 Trawls and dredges

Trawls and dredges are often called towed gear or dragged gear.

Catching principle

The catching principle and construction of trawls are shown in Figure 7.

Construction

Trawls and dredges are in principle netting bags that are towed through the water to catch different target species in their path. During fishing, the trawl entrance or trawl opening must be kept open. With beam trawls and dredges this is done by mounting the trawl bag on a rigid frame or beam. With otter trawls the opening is maintained by so-called otter boards (trawl doors) in front of the trawl which keep the trawl open sideways while the vertical opening is maintained by weights on the lower part (ground-rope) and floats on the upper part (headline). With pair trawling, the vertical opening is also maintained by weights and floats, while the lateral opening is maintained by the distance between the two vessels that are towing the trawl.

In otter trawling, the trawl is connected to the trawl boards by a pair of sweeps (rope or steel wire) and the trawl doors are connected to the vessel by a pair of warps (normally steel wire). In otter trawling and partially in pair trawling, the sweeps and warps are also part of the catching system, as they will herd fish towards the centre of the trawl path and the approaching trawl, so that the trawl may catch fish over a larger area than that of the trawl opening. With beam trawl and dredges there is little or no herding of target species in front of the trawl, so the effective catching area is that of the trawl or dredge opening.

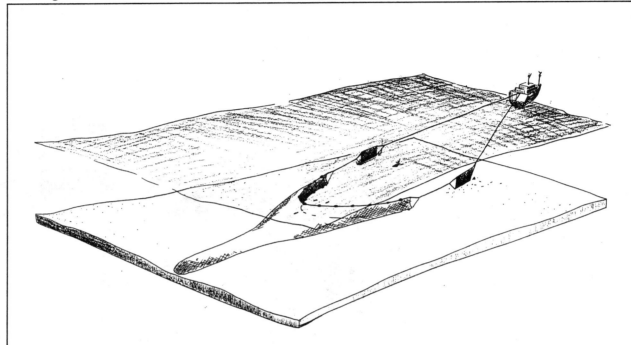

Figure 7. Catching principle and construction of an otter trawl, showing the trawl connected by the sweeps to the trawl doors (otter boards) and the warps between the trawl doors and the vessel.

Operation

Beam trawls and dredges are exclusively operated on the bottom, where they are towed for a certain length of time (towing time) and distance before being retrieved for the emptying of the catch and being set again for another tow.

Otter trawls and pair trawls are most often operated on the bottom to catch different demersal target species. However, these gears are also commonly used for pelagic (or mid-water) trawling at different depths between the surface and the sea bed. This is done by attaching more floats to the head rope of the trawl opening as well as regulating the trawl depth by varying the length of warp and towing speed. In most pelagic trawling, the trawl depth is monitored by depth sensors on the trawl, so that the fishing depth can easily be adjusted to that of the fish targets.

Target species

Beam trawls are mainly used for catching flatfishes such as plaice and sole as well as for different species of shrimp. Dredges are commonly used for harvesting scallops, clams and mussels. Demersal otter and pair trawls are used to catch a great variety of target species like cod, haddock, hake, sandeel, flatfish, weakfish, croakers as well as shrimps. Pelagic trawls are used in the fisheries for various pelagic target species, like herring, mackerel, horse-mackerel, blue whiting and pollock.

4.3 Seine nets

Catching principle

The catching principle and construction of seine nets is shown in Figure 8.

Seine netting (including two variations known as Danish seining and Scottish seining) can be described as a combination of trawling and seining (see below). When setting the gear, the first warp (rope) is attached to an anchor with a surface buoy (Danish seining) or a buoy only (Scottish seining) and set in a semicircle. Then the seine bag is set before paying out the second

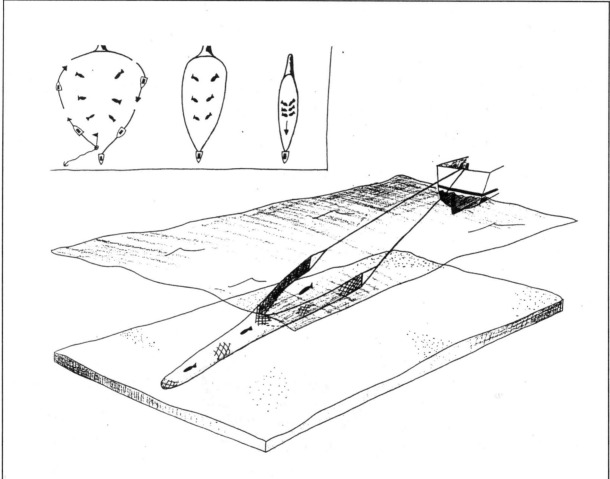

Figure 8. Catching principle and construction of seine nets. Expanded view: three stages of the catching process.

warp in another semicircle back to the buoy (attached to the anchor in Danish seining). When the seine and warps have sunk to the bottom, the warps are hauled. As they are tightened, the warps move inwards towards the centre line between the vessel and the seine bag. Fish in the encircled area will then be herded towards the central part of the area. As the warps are further tightened, the seine bag moves forward and catches the fish.

Construction

As mentioned above, the main parts of a seine net are the seine bag and the warps. The seine bag is similar to a trawl bag, where the entrance is kept open by floats on the headline and a weighted ground line (foot rope). The warps are usually made of heavy rope, so that they maintain good contact with the bottom for as long as possible during tightening in order to herd the fish towards the central area for later capture by the seine bag.

Operation

The seine net was originally constructed for the capture of flatfish on soft and smooth bottoms and operated as described above. In later years this gear has also been developed to be operated on rougher bottoms and in the pelagic zone. A more recent mode of operation is for instance used on mid-water shoals of cod. The fishing depth of the seine net is then determined by large surface floats connected to the head rope of the seine net by lines, the length of which corresponds to the desired fishing depth.

Target species

The seine net is still commonly used to catch different flatfishes such as plaice and sole, but has in recent years become an important gear also for cod and other demersal target species.

4.4 Beach seines

The catching principle and construction of beach seines are shown in Figure 9.

Catching principle:

The operation of beach seines is based on encircling fish schools by a netting wall, made of webbing where the meshes are so small that the target species does not get entangled.

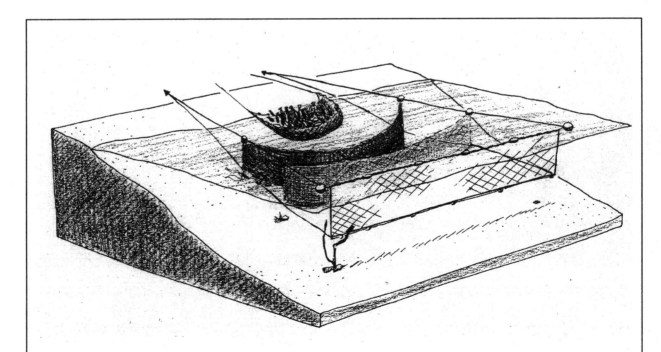

Figure 9. Catching principle and construction of (beach) seines, showing four stages of the catching process.

Construction

The beach seine is an ancient gear that is still widely used. The seine consists of a wall of webbing, e.g. with a depth of 5m by 100m length, with an upper float line and a lower sinker line. In principle a similar construction to the gillnet, but with smaller meshes so that the fish is entrapped instead of being gilled or entangled. At either end of the seine there are long warps (rope).

Operation

As the name indicates, the beach seine is operated from the beach – using the beach as an additional barrier in the catching process. The gear is normally operated from a small vessel. First, one of the end warps is paid out perpendicular to the beach. Then the seine is set parallel to the beach and the second end warp is taken back to the beach. The warps are pulled in so that the seine approaches the beach in a semicircular form – and most of the fish in the area between the seine and the beach are likely to be caught. In many seine fisheries, both beach seine and purse seine, light is used to attract and concentrate fish before the seine is set.

Target species

Beach seines catch a variety of inshore fish species, both demersal and pelagic.

4.5 Purse seines

Catching principle

The purse seine is used to encircle fish schools in mid-water, close to the surface, by a netting wall with small meshes. The lower part of the net is then closed to prevent escapement by diving.

Construction

The purse seine was developed in the 20th century, for offshore fishing. Basically its construction is similar to that of the beach seine. However, below the sinker line, the purse seine is equipped with a series of metal purse rings spaced at regular intervals. By hauling in the purse line that runs through the purse rings, it is possible to purse and close the bottom part of the seine so that the encircled fish cannot escape.

Operation

The purse seine is always operated from a vessel, varying in size from small coastal purse seiners of 15m in length to large ocean going purse seiners as large as 100m length. When a fish school has been located, the catching operation starts by dropping a surface buoy with a line connected to the end of the seine. As the vessel moves forward, the drag from the buoy line will pull the purse seine overboard and the seine is paid out in a circle around the fish school. When the setting is completed, the buoy is picked up, and the purse line is pulled in to purse and close the bottom of the seine. Then the seine is pulled in until the fish are concentrated in the last (and often reinforced) part of the seine, from where it is taken on board with a brailer (large dip net) or by a fish pump.

In modern purse seine fisheries, hydro-acoustic equipment (sonar) is widely used for locating fish schools and also for monitoring the position of the school relative to the gear during the setting of the seine.

Target species

Purse seine fishing is used almost exclusively for pelagic fish like herrings, sardines, sardinellas, anchovies, mackerels and tunas.

4.6 Other fishing gears and devices

The most common fishing gears and methods are briefly described above. There are, however a great variety of different gears which are specialized varieties of the main gear types and methods. For more details or description of gears that falls beyond the scope of this manual, it is recommended to use some of the more comprehensive works on fishing gears and methods cited under Section 7 "Recommended Reading". Some devices and techniques are, however, used with different fishing gears to improve the catch efficiency:

Light is used to attract fish in many fisheries, but is most often used with purse seining, beach seining or other varieties of seine fishing. In darkness the light will either attract the target species directly or indirectly by attracting and illuminating prey organisms. The light source is often lamps mounted on a vessel or small raft. After some time the net is set around the light source with the attracted fish.

FADs (fish aggregating devices) are also commonly used in some areas to aggregate fish. These can consist of anchored rafts of logs or other material and act as artificial habitats that will attract fish and other organisms over time and hence create good fishing spots that can be exploited by different gears. This is also the case with artificial reefs that either are made by shipwrecks or by deliberately dumping objects on otherwise flat and sandy sea beds to create fish habitats. The Code of Conduct (Paragraphs 8.11.1 – 8.11.4) encourages the use of such structures providing they are used in a responsible manner.

Different stupefying devices are also used to capture fish. Explosives (dynamite) will stupefy fish, some of which will float to the surface so that they can be picked up. The use of explosives with fishing is, however, regarded as a very destructive practise as the explosion most often kills much more fish than those that are caught – and can in addition ruin valuable fish habitats like coral reefs.

Different chemicals can also be used in the same way. Rotenone (a poison derived from plants) is one of the best known examples used to stupefy fish, mainly in fresh water systems. As with explosives, the use of chemicals gives a high risk of killing much more fish and other organisms than those that are harvested. The use of chemicals with fishing should therefore be considered as not responsible.

The Code of Conduct (Paragraph 8.4.2) specifically calls for the prohibition of "dynamiting, poisoning and other comparable destructive fishing practices".

5. GEAR SELECTIVITY AND ECOSYSTEM EFFECTS OF FISHING

Before commenting on the selectivity properties and ecosystem effects of different fishing gears it may be useful to give a brief description of relevant factors and definitions.

The catching process

The catching process starts as the fishing gear is deployed in the water and ends as it is retrieved from the water, be it ashore or on the deck of a fishing vessel. Throughout the catching process there may be encounters between the gear and various fish and other marine organisms, including sea birds, and bottom habitats.

Ecosystem effects of fishing

The effect of fishing on the ecosystem is primarily the removal of the organisms caught in the fishery, but also includes the direct and indirect effects caused by the gear during the catching process – like destruction of bottom habitats (e.g. corals), "ghost fishing" by lost fishing gears, pollution etc.

Selectivity

The selectivity of a certain fishing method depends on its ability to select the desired ("target") species and sizes of fish from the variety of organisms present in the area where the fishery is conducted.

The total selectivity of a fishing method is the combined result of the inherent selective properties of the fishing gear and the way it is operated. With most fishing gears it is possible to impair or improve the selectivity by changing the gear configuration or the operation. For example, in trawl fishing the catch of small fish can be reduced by increasing the mesh size and/or by the use of sorting devices like sorting grids or large mesh panels that allow for escapement of the smaller fish (See Figure 10). The fisher can also select for target species and sizes by avoiding areas and periods where there is a high probability of catching small fish or otherwise undesired bycatch. The Code of Conduct requires the minimisation of the catch of non-target species and of discards (Paragraph 7.2.2 and Sub-article 8.5).

Bycatch

Bycatch is anything that is caught in the fishing process beyond the species and sizes of the targeted marine organisms. There is a great variety of bycatch species, ranging from sponges and corals to unwanted or unmarketable fish species or sizes, as well as turtles, marine mammals and sea birds. Bycatch can be classified into three main groups: marketable and legal; non-marketable; and/or non-legal. Non-economic bycatches consist of organisms that are non-marketable for the fisher, while non-legal bycatches are sizes or species of marine organisms that are protected by regulations.

Thus, marketable and legal bycatch is welcomed by the fisher, while all other bycatch should be avoided. However, the capture of some bycatch is normally unavoidable. Most fisheries regulations do therefore allow for a certain amount of bycatch, e.g. a certain percentage of undersized catch of a target species or a certain amount of an otherwise protected species. For example, in the New Zealand trawl fishery for squid, a certain number of sea lions are tolerated as bycatch, but the fishery is closed as soon as this number is reached for the specific fishing season. In the Barents Sea cod fishery, a bycatch of 15% (in numbers) of undersized fish (less than 42 cm) is tolerated within the legal frames of the fishing regulations.

Discards

Discarding, which means to throw parts of the catch back into the water is a common practice in most fisheries, although the amount of discard varies significantly between different fisheries. Discards are most often organisms that are not marketable or give a low price compared with the more valued target species. The survival of discarded organisms depends on the ability to survive in air, the time they are kept out of the water and how they are handled before being discarded. However, it should be expected that most discarded organisms suffer high mortality, which adds a "hidden mortality" to the fishing mortality that is calculated from the landed catch.

Unintentionally, modern fisheries management has encouraged increased discarding in many fisheries. With the introduction of quotas and licences for different species it is often illegal to catch and land certain species of fish. Thus it is not uncommon, particularly in mixed-species fisheries, to find large-scale discarding of valuable, marketable fish (e.g. cod and saithe) because the trip quota or monthly quota has been exceeded. In addition, in some fisheries, small fish under the legal size limit for landings may also be discarded.

"High grading" is another form of discarding where only the most profitable part of the catch is retained, while less valuable fish are discarded. This phenomenon is also often linked to a quota system where the fisher tries to get the maximum value from a limited quota by keeping only the most valuable part of the catch and discarding the rest.

By-mortality

By-mortality is the mortality of marine organisms from injures caused by encounter with the fishing gear during the fishing process. One example is fish that die from infections or osmotic imbalance caused by scale loss after escapement through trawl or gillnet meshes. The introduction of mesh size regulations e.g. in trawls should therefore take into account, and be accompanied by studies of, the survival of fish that are released through trawl meshes or sorting devices. If the released fish suffer high mortality there is little to be gained by sorting it out of the fishing gear, which is the case with vulnerable species like herring. On the other hand, studies have shown that cod and several other demersal species have high survival rate after escapement from or encounter with fishing gear.

Ghost fishing

The term "ghost fishing" is used to describe the capture of marine organisms by lost or abandoned fishing gear. This is particularly a problem with gillnets, trammel nets and pots. The gear is usually lost because it becomes stuck on rough bottoms containing corals and stones, causing the buoy line to break during retrieval. Nets or pots may then continue to fish for years. Captured fish and crustaceans will die and serve as attracting bait for more fish and other organisms. Ghost fishing may therefore represent a serious problem in many areas, causing "hidden fishing mortality" over a long period of time. Paragraphs 7.2.2 and 8.4.6 of the Code of Conduct draw attention to the need to minimise ghost fishing.

Habitat effects

Destruction of bottom habitats is particularly a problem with the use of dragged demersal gear like beam trawls, otter trawls and dredges. Corals and other epifauna have been and may be destroyed over large areas. It is still debated as to whether these gears have any real negative effect on soft, sandy bottoms. However, it has been documented that trawling has ruined large areas of coral, which has a very low recovery rate, and other epifaunal organisms. Here, the Code of Conduct calls for the development and use of environmentally safe fishing gear (Paragraph 7.2.2).

Catch quality

Properties of fishing gears and the way in which they are operated also affect the quality of the catch, thus having an indirect ecosystem effect by the misuse of natural resources. In gillnet fishing, poor quality results from too long a soak time. This results in fish dying in the nets and either rotting or becoming damaged by scavengers; therefore, that part of the catch is not marketable and has to be discarded. This may also be a problem in longline and pot fishing. In trawling, particularly with large catches, it is not uncommon that part of the catch is ruined by squeezing in the trawl bag or becomes of inferior quality because of too long storage on deck before it is processed. This too is contrary to the requirements of the Code of Conduct (Paragraph 8.4.4).

Energy efficiency

Use of energy, particularly fossil fuel, is also an ecosystem related aspect of fisheries. The energy efficiency (i.e. fuel consumption per unit of landed catch) varies considerably with different fishing gears and methods, from negligible use of fuel to more than 1 litre of fuel per kilogram of landed catch. This is covered in Sub-article 8.6 of the Code of Conduct, calling for energy optimization.

Pollution

Fisheries can contribute to air pollution through emission of combustion gases. The relative pollution effect from different fisheries is closely related to their energy efficiency.

Pollution of water from fisheries is mainly by loss of fishing gear or by deliberately discarding old gear and equipment as well as oil products and chemicals at sea. These two aspects are covered by Sub-articles 8.8 and 8.7 respectively of the Code of Conduct.

5.1 Selectivity properties and ecosystem effects of different fishing methods

The following is a description of the general selectivity properties and ecosystem effects of the different fishing gears and methods mentioned under Sections 3 and 4. For a proper evaluation of the ecosystem effects of fishing, each specific fishery has to be analysed separately, as the selectivity properties and ecosystem effects of a certain fishing method may vary considerably with geographical area, time of year and how the gear is operated.

A few examples are the following.

- Gillnets that are operated in shallow water and are hauled on a daily basis will yield catches of higher quality and with less risk of gear loss and ghost fishing compared with gillnets operated in deep water with a soak time of several days.

- Pelagic gillnets that are fished in the vicinity of sea bird breeding sites may have high bycatches of sea birds in the breeding season, but not in other periods.

- Bycatch of juvenile fish in demersal trawling may vary considerably with the availability of juveniles, the species composition, towing speed and general catch rates.

So even if the intrinsic selective properties and other ecosystem impacts of a certain fishing gear may be regarded as fairly constant, the effects caused by the gear on fish stocks and the rest of the ecosystem may vary with diurnal, seasonal and long term changes in species and size composition of organisms available to the fishing gear and with differences in fishing practice.

(a) Gillnets

In general gillnets are considered to be very size selective, with catches of fish sizes that correspond well to the chosen mesh size. However, due to entangling a small proportion of . larger and smaller fish may be taken. The species selectivity of gillnets is not particularly good and as different fish species grow to different sizes, there is always a possibility of catching juveniles of a large species when using small mesh gillnets for a smaller target species. Another negative impact of gillnets is the bycatch of sea birds, marine mammals and turtles. Although little information exists on the real effect of such bycatches on the populations of these organisms, it has generated concerns, particularly for pelagic gillnet fishing.

Information on by-mortality of fish after escapement from gillnets is scarce. However, observations of fish with wounds from gillnet meshes are commonly made in catches by other gears, but the actual mortality rates from such injuries are not known.

Ghost fishing is one of the most criticized aspects of gillnet fishing, and may have severe negative effects, particularly in deep water gillnet fisheries. The energy efficiency of gillnet fishing is generally high with a correspondingly low air pollution effect.

As mentioned above, the catch quality of gillnet caught fish can be high. However, gillnets that are operated with soak times of several days tend to produce catches of inferior quality, as fish caught early in the fishing period may die and start to deteriorate long before the nets are retrieved.

(b) Trammel nets

Compared with gillnets, trammel nets have very poor size selective properties and they will also catch a greater variety of species. The problem of ghost fishing is, however, lower as trammel nets generally are operated in shallow water with less risk of gear loss, but ghost fishing problems should nevertheless be anticipated due to loss of trammel nets that get stuck on rough bottoms like coral reefs.

(c) Handlining and trolling

Handlining and trolling are not particularly size selective and in principle not very species selective either. However, these gears are commonly used in specific seasons or at specific grounds where the fishers, by experience, are able to catch only one or a few species, so that the catches are usually dominated by a few targeted species. Otherwise, handlining and trolling are generally regarded as ecosystem-friendly ways of fishing which produce catches of high quality.

(d) Longlining

Despite the fact that longlines may attract and catch a large variety of fish species and sizes, this gear is considered to have medium to good species and size selective properties. The species selectivity of longlines can clearly be affected by the type of bait used, as different species have been shown to have different bait preferences. The size selective properties can partly be regulated by the hook and bait size as many studies have shown a correlation between the size of hook and bait and the size of the fish caught. The longline attracts fish from several hundred meters away, and as large fish have a greater swimming and feeding range than smaller fish, this adds to the size selective properties of longlines.

Bycatch of marine mammals is no particular problem with longlining, but there might be significant bycatches of different seabirds, which mainly are caught as they try to catch the baited hooks during the setting of the lines. This problem has been recognised by FAO member States, leading to the development of the FAO International Plan of Action (IPOA) for Reducing

Incidental Catch of Seabirds in Longline Fisheries[1]. The IPOA specifies some optional technical and operational measures for reducing the incidental catch of seabirds, including, for example, increasing the sinking rate of baits and the use of bird scaring lines which are towed behind the vessel, above the longline being set.

Little is known about the by-mortality of fish in longline fishing, but fish that are lost during retrieval of longlines do often suffer mortality. Ghost fishing may be regarded as a minor problem with longlining and this gear is not considered to cause significant adverse habitat effects. The energy efficiency of longlining is generally high, with typical energy coefficients from 0,1 to 0,3 (kilogram fuel per kilogram of landed catch), which is in the same range as that of gillnetting.

Longline caught fish are in general of high quality, but as is the case for gillnetting, long soak times may lead to reduced catch quality mainly due to bottom scavengers (sea lice, hagfish) that may attack and eat parts of the hooked fish.

(e) Pots

As with longlines, the species selectivity of pots may be regulated by the bait used. Lobster fishermen do for instance often use "sour" or rotten fish as bait to avoid catching crabs in their lobster pots. As with longlines, the attraction of fish and crustaceans to baited pots tends to attract the larger animals in the fished area. The size selectivity of pots may be further improved by the use of so called escape gaps, the size of which allows for escapement of smaller animals. By-mortality is not regarded as a problem with pot fishing and this gear has negligible effect on bottom habitats. There is, however, a certain risk of ghost fishing, as lost pots may continue to fish long after they are lost. This can be reduced by having certain parts of the pot made from a bio-degradable material. Furthermore, pot fishing is regarded to have a high energy efficiency and good to superior catch quality, as the catch normally remains alive and in good condition.

(f) Traps

Traps are usually constructed of relatively fine meshed webbing to avoid the tangling of fish and other organisms, so their size and species selectivity is generally low. As the caught animals usually stay alive, and as traps are most often operated in shallow water, they allow for easy release and high survival of unwanted catch organisms. With responsible fishing practices, the actual selectivity properties of traps may therefore be good and the by-mortality is low. Traps in general have little adverse impact on bottom habitats, they do not create ghost fishing problems and the energy efficiency and catch quality of trap fishing are high.

(g) Spears and harpoons

The capture of fish and other animals by spears and harpoons is probably one of the most environmentally friendly fishing methods. As the target is identified before capture, the fisher can be very selective regarding both species and size of prey. Fishing with spear or harpoon may give some by-mortality of wounded animals that escape and, when used in reef areas, the use of spears can lead to damage of coral, but apart from these, there are no substantive adverse effects related to ghost fishing or habitat destruction and the energy efficiency and catch quality is high.

---

[1] See http://www.fao.org/fi/ipa/incide.asp for the full text of the Plan of Action.

(h) Pelagic trawls

Pelagic trawls generally have high species selectivity as they are commonly used for catching schooling pelagic fish that tend to occur in single-species aggregations. The size selectivity is poorer as the fish bag of the trawl is usually made from small mesh webbing to avoid meshing by smaller individuals. Successful trials have been done with sorting grids that effectively release the smallest fish (e.g. with trawling for mackerel). However, these have not been applied in practical fishing, as many pelagic fish seem to suffer high mortality after being released from the fishing gear – mainly caused by the loss of scales which easily leads to secondary infections and osmotic imbalance. At present, sorting systems for the release and protection of juvenile pelagic fish is therefore not recommended. By-mortality is hence a minor problem with pelagic trawling and this gear, naturally, does not have any ghost fishing or habitat destructive effects.

The fuel consumption of pelagic trawling can be high, but still the energy efficiency might be relatively good as large catches are often made during short time periods. The catch quality of pelagic trawls is also relatively high, although large catches may give some squeeze and pressure damage to the fish in the trawl.

(i) Demersal trawls

Demersal trawls are used for the capture of a great variety of bottom fish and crustaceans (mainly shrimps and prawns) while dredges are used to harvest molluscs (clams and scallops).

(i1) Fish trawls

Otter trawls are widely used for the capture of different demersal fish species – most often in so-called mixed-species fisheries. The size selectivity may to a certain degree be regulated by the cod-end mesh size. Ideally, a certain mesh size should allow for the release of all fish below a certain size. However, the mesh size selection of trawls may be hampered in many ways. With increasing catch load in the cod-end, the meshes tend to stretch and close, so that the effective mesh size is significantly reduced. Clogging (closing) of meshes by fish getting stuck in them is another common problem that leads to poorer selectivity. The choice of mesh size for a certain target species may not give ideal selection of other species with different growth characteristics, a problem that is related to all mixed-species fisheries.

Considerable research and development effort has been spent in recent years to improve the size and species selectivity of trawl gear. Different solutions have been developed and implemented, like the sorting grid that is now used in many demersal trawl fisheries. Figure 10 illustrates the sorting grid which is now mandatory in the Barents Sea bottom trawl fishery for cod and other demersal species. Most of the juvenile fish will escape through the openings between the metal bars of the grid, while larger fish will be retained.

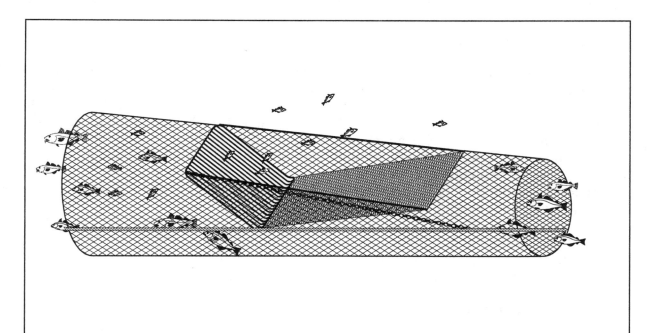

Figure 10. Section of fish trawl with sorting grid. Smaller fish are released through the grid slots, while larger fish swim or are swept down under the grid and go to the cod end.

Where introduced, sorting grids and other selective devices such as larger square mesh panels and escape gaps have lead to improved selectivity of bottom trawls, first of all the size selectivity, but also the species selectivity, although there is still a need for further improvements.

Releasing juvenile fish has little benefit if they do not survive. Extensive studies that have been carried out on demersal target species like cod and haddock have shown very low by-mortality of fish that have escaped through meshes or sorting grids. Although studies of survival after escapement so far have only been done for a restricted number of species, there seems to be a general indication of high survival of demersal fish after encounters with and escapement from fishing gears, when the fish are sorted out and released at fishing depth.

Demersal trawls do inevitably have an effect on bottom habitats. Several studies have been carried out within this field, however, without any general conclusions. On soft and sandy bottoms, there might be an inverse effect on the species composition in an area with a shift towards species that are less dependent on the epifauna removed by trawling. Most studies do, however, indicate that soft bottom habitats will be restored after some years without trawling

On hard bottom, trawling is likely to cause more long lasting or irreversible habitat effects, e.g. by destroying corals that have restoration periods from decades to more than a hundred years. Large areas of coral bed have already been destroyed by bottom trawling, particularly with the development of heavier and stronger trawl gear.

Trawls are occasionally lost, but this gear loss is not associated with any risk of ghost fishing.

The energy efficiency of demersal trawling is low and air pollution from the emission of exhaust gases is correspondingly high due to the high energy needed for pulling the net, doors, sweeps and warps through the water.

The catch quality of trawl caught fish varies with the amount of catch and the towing time. Large catches do often lead to lower catch quality because of the squeezing of the fish in the trawl bag and a longer time before the last part of the catch is processed on board.

(i2) Shrimp trawling

Shrimp trawls are in principle comparable to demersal otter trawls for finfish capture and also have comparable ecosystem effects, except that the selective properties of shrimp trawling are very poor. This is due to the small meshes that have to be used in shrimp trawls in order to retain these relatively small target species. Shrimp trawling does therefore produce relatively large amounts of bycatch and a high proportion of this is discarded. The development of sorting grids has, however, improved the species- and size selectivity in many shrimp trawl fisheries, as most fish over a certain size are released from the trawl through the sorting grid or bycatch reduction device (Figure 11). Bycatch of the youngest fish groups (1-2 year olds) is still a problem, as they have overlapping sizes with those of the shrimps.

Bycatch of turtles during shrimp trawling operations is a problem in some areas but is being widely addressed through the use of Turtle Excluder Devices (TEDs) which operate on similar principles to other bycatch reduction devices.

(i3) Beam trawling

The ecosystem effects of beam trawling are to a large extent comparable to those of demersal otter trawls. Compared with otter trawls, however, beam trawls will generally have a poorer energy efficiency and a stronger impact on bottom habitats.

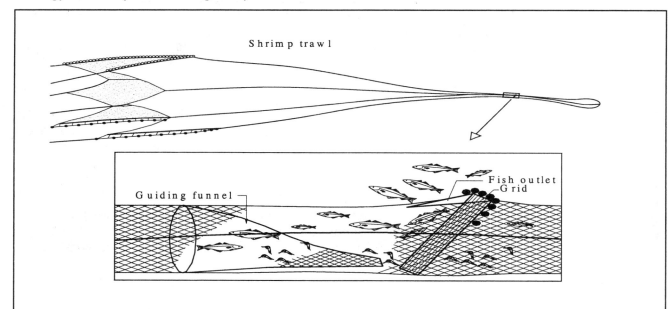

Figure 11. Shrimp trawl with sorting grid (expanded view). Shrimp and fish that pass backwards in the trawl are guided by a funnel to the bottom of the backwards slanting metal grid. Shrimp and fish of comparable size will pass through the slots of the grid and go to the cod end, while larger fish and other organisms (e.g. jellyfish) will slide upwards over the grid and are released through the outlet.

(j) Seine nets

Seine netting is also fairly similar to demersal trawling with respect to most ecosystem effects. However, the seine net is considered to cause less habitat destruction, is more energy efficient and does produce a better catch quality overall.

(k) Purse seine

Purse seining is a non-selective gear regarding fish size, as the mesh size is chosen to be so small that there should be no risk of mass meshing of fish, even by the smallest size groups of the target species. However, in cases where the fish size in the catch is too small, as estimated from samples taken from the seine, there is usually an opportunity to release the fish. The species selectivity is fairly high and both from the fishers experience and by use of modern sonar equipment it is not too difficult to identify the species before the seine is set.

There is a certain risk of by-mortality in purse seining. Pelagic fishes are in general sensitive to contact with fishing gears which easily leads to loss of scales and resulting mortality. This can be related to the above-mentioned release of unwanted species or sizes of fish, but the main cause of by-mortality in purse seining is the escapement of fish after net rupture due to large catches and/or bad weather.

There is extremely low risk of ghost fishing with lost purse seines. The energy efficiency is high because of the relatively large catches that give a high catch-per-unit-effort in this fishery. Catch quality is normally also high, particularly in modern purse seining, where the catch is pumped directly into refrigerated tanks on the fishing vessel.

Purse seining has generated some adverse publicity as a result of bycatches of dolphin in some tuna fisheries, but effective methods to avoid such capture have been developed.

(l) Beach seines

Beach seines have poor selectivity properties, catching a variety of species and sizes of fish and other organisms. There may be some by-mortality associated with beach seining, while the energy efficiency and the catch quality of this gear is generally high.

6. MANAGEMENT CONSIDERATIONS: SELECTIVITY AND OTHER ECOSYSTEM EFFECTS OF FISHING

Table 1 gives an example of how the properties of different fishing gears could be evaluated in terms of their selectivity and ecosystem effects. Here the various ecosystem effects are given a rank from 1 (non-favourable) to 10 (favourable), giving an overall index of average ecosystem effect. This table must of course only be regarded as a guideline and an example of how to approach an evaluation of different fishing gears and fisheries from a management point of view. Thereafter the specific fishery in an area should be analysed in more detail and the proposed ecosystem factors should also be weighted according to their importance in a local or regional case. Although a fishing method might be characterised as being more or less responsible in general, attention should be paid to where, when and how it is being used. The evaluation and elaboration of technical regulations should be done in co-operation with the fishers to establish better understanding by them of the aim of the regulations and to hear and consider their advice on the regulations and their implementation (Chapters 7 and 8).

Table 1. Generalized estimate of ecosystem effects of fishing for different fishing methods – ranked on a scale from 1 (non-favourable) to 10 (highly favourable) with respect to different ecosystem related factors.

Ecosystem effects and Gear type	Size selection	Species selection	By-mortality	Ghost fishing	Habitat effects	Energy efficiency	Catch quality	Ecosystem effect index
Gillnets	8	4	5	1	7	8	5	5,4
Trammel nets	2	3	5	3	7	8	5	4,7
Handlining	4	4	6	10	9	9	9	7,3
Longlining	6	5	6	9	8	8	8	7,1
Pots	7	7	9	3	8	8	9	7,3
Traps	5	5	8	8	9	9	9	7,6
Spear, harpoon	8	9	5	10	10	8	9	8,4
Pelagic trawl	4	7	3	9	9	4	8	6,3
Demersal trawl	4	4	6	9	2	2	6	4,7
Beam trawl	4	4	6	9	2	1	6	4,6
Shrimp trawl	1	1	7	9	4	2	6	4,3
Seine net	5	5	6	9	4	5	8	6,0
Purse seine	-	7	5	9	9	8	8	7,7
Beach seine	2	2	5	10	6	9	9	6.1

Other specific factors, for example their socio-economic implications, could be added to a specific evaluation, as a guideline to future management strategies with respect to choice and priorities between different fishing methods. These factors should be included with the information used to assist in the design of management strategies (Chapter 5).

This form of evaluation can also be used to identify present or future ecosystem-related weaknesses with existing fishing methods and practises as a baseline for research and development aiming at improvements of selective properties and undesired ecosystem effects.

7. RECOMMENDED READING

Alverson, D.L., Freeberg, M.H., Murawski, S.A. and Pope, J. 1994. A global assessment of fisheries bycatch and discards. *FAO Fisheries Technical Paper* **339**. FAO, Rome. 233pp.

Ben-Yami, M. 1989. *Fishing with Light (An FAO Fishing Manual)*. Blackwell Science Ltd., Oxford. 132pp.

Bjordal, Å. and Løkkeborg, S. 1996. *Longlining*. Fishing News Books, Blackwell Science Ltd., Oxford. 156 pp.

Brandt, A. von 1984. *Fish Catching Methods of the World*. Fishing News Books, Farnham.

(See Gabriel for revised edition of this work)

Cowx, I.G. and Lamarque, P. (eds.). 1990. *Fishing with Electricity (Applications in freshwater fisheries management)*. Blackwell Science Ltd., Oxford. 272 pp.

Fernø, A. and Olsen, S. (eds.). 1994. *Marine Fish Behaviour in Capture and Abundance Estimation*. Fishing News Books, Blackwell Science Ltd., Oxford. 221pp.

Gabriel, O. (ed.). (in prep.). *Fish Catching Methods of the World. (Fourth Edition)*. Blackwell Science Ltd., Oxford. 448 pp.

Hall, S. 1999. *The Effects of Fishing on Marine Ecosystems and Communities*. Fishing News Books, Blackwell Science Ltd., Oxford. 296pp.

Kaiser, M.J. and de Groot, S.J. (eds.). 2000. *The Effects of Fishing on Non-target Species and Habitats*. Blackwell Science Ltd., Oxford. 399pp.

Moore, G. and Jennings, S. 2000. *Commercial Fishing (the wider ecological impacts)*. Blackwell Science Ltd., Oxford. 72pp.

Nédélec, C. and Prado, J. (eds.). 1989. *FAO Catalogue of Small Scale Fishing Gear*. Blackwell Science Ltd., Oxford. 224pp.

Nédélec, C. and Prado, J. 1990. *Definition and classification of fishing gear categories. FAO Fisheries Technical Paper* **222**. Revision 1. FAO, Rome. 92pp.

Nolan, C.P. (ed.). 1999. Proceedings of the International Conference on Integrated Fisheries Monitoring. Sydney, Australia, 1-5 February 1999. FAO, Rome. 378pp.

Robertson, J. (in prep.) *Minimising Discards to Improve Global Fish Stocks*. Blackwell Science Ltd., Oxford. 256pp.

Sainsbury, J.C. 1996. *Commercial Fishing Methods (an introduction to vessel and gear)*. Blackwell Science Ltd., Oxford. 368pp.

Scharfe, J. (ed.). 1989. *FAO Catalogue of Fishing Gear Designs*. Blackwell Science Ltd., Oxford. 160pp.

CHAPTER 3

THE USE OF TECHNICAL MEASURES IN RESPONSIBLE FISHERIES: AREA AND TIME RESTRICTIONS

by

Stephen HALL

Australian Institute of Marine Science, Townsville, Australia

1. WHAT ARE AREA AND TIME RESTRICTIONS? 49
2. WHY WOULD YOU ESTABLISH AREA OR TIME RESTRICTIONS?.....................................51
   2.1 As a fishery management measure..51
   2.2 As a wider conservation measure 53
   2.3 To resolve equity issues...54
3. WHAT ARE THE ADVANTAGES AND DISADVANTAGES OF AREA AND TIME RESTRICTIONS? 55
   3.1 Advantages..55
   3.2 Disadvantages 56
4. CASE STUDIES 57
   4.1 Gulf of Mexico: a mixture of area and time closures 57
   4.2 Coral reefs...57
   4.3 Surf zone fisheries 60
   4.4 Georges Bank..60
   4.5 The Plaice Box 61
   4.6 Adaptive management using the closed area approach: an example for NW Australia 61
5. WHAT ARE THE PRACTICAL STEPS TOWARDS ESTABLISHING TIME AND AREA RESTRICTIONS? 66
   5.1 Set your goal 66
   5.2 Specify criteria for selection 66
   5.3 Assemble information and conduct a preliminary evaluation 69
   5.4 Implement the process of negotiation 71
   5.5 Evaluate the need for underpinning research 72
6. CONCLUDING COMMENTS 73
7. REFERENCES 73

1. WHAT ARE AREA AND TIME RESTRICTIONS?

This chapter describes approaches to fisheries management that restrict access by fishers to an area in some way. In some cases restrictions are imposed throughout the year, whereas in others they only apply at a particular time (or times), usually in certain seasons. As noted in Chapter 4, when an area or time restriction is established as a technical conservation measure it is a form of input control. There are, however, many other objectives beyond stock conservation that can be served, particularly by permanent area closures, thereby warranting a separate treatment of this topic.

These more general objectives are articulated under Article 2, Paragraph g of the FAO Code of Conduct for Responsible Fisheries, which states that fisheries should "promote protection of living aquatic resources and their environments and coastal areas".

Area closures (whether temporary, seasonal or permanent) are referred to by a variety of names, each of which may have a particular formal definition, depending on the legislative or cultural context. Of these various terms, however, 'Marine Protected Area' or MPA is, perhaps, the most widely used. The International Union for Conservation of Nature (IUCN) defines an MPA as: *"Any area of intertidal or subtidal terrain, together with its overlying water and associated flora, fauna, historical and cultural features, which has been reserved by law or other effective means to protect part or all of the enclosed environment"*. Kelleher & Kenchington (1992).

Similarly, in the Canadian Oceans Act, a Marine Protected Area is defined as:

an area of the sea...(that) has been designated ... for special protection for one or more of the following reasons:

(a) the conservation and protection of commercial and non commercial fishery resources, including marine mammals, and their habitats;

(b) the conservation and protection of endangered or threatened marine species, and their habitats;

(c) the conservation and protection of unique habitats;

(d) the conservation and protection of marine areas of high biodiversity or biological productivity; and

(e) the conservation and protection of any other marine resource or habitat as is necessary to fulfill the mandate of the Minister (of Fisheries and Oceans).

Section 35(1) Oceans Act, Canada.

Note that both of these definitions leave considerable latitude with respect to the nature of the restrictions on fishers that are imposed and a number of classification schemes have been proposed for categorizing the various levels of area restriction that would fall under the MPA banner. As with most such schemes, however, there are shades of grey at the boundaries between classes and it may be better to think in terms of a continuum between absolute prohibition of access at one extreme (often termed a 'no-take' reserve) and relatively minor restrictions such as limitations on gear that are deemed to be threatening to key environmental or conservation values at the other.

One important distinction that needs to be made, however, is between area closures that are established solely with reference to fishing activity and multiple-use management areas, which allow a range of activities to take place, but with appropriate restrictions on them to protect valued attributes of the area. On coral reefs for example, tourist operators might be restricted to certain mooring locations to limit anchor damage and fishers might have restrictions placed on the types of gear that can be used.

Such multiple-use zones are most likely to be established if a coastal management framework exists that seeks to reconcile the requirements and values of all legitimate interested parties. Thus, the consultative and legislative context in which multiple-use areas are established and managed is likely to be markedly different from that which obtains when an area is established solely within a fisheries management domain. While much of this chapter will deal with the fishery specific issues surrounding closed areas, the reader should bear in mind the wider set of considerations that must be factored in for multiple-use management areas.

2. WHY WOULD YOU ESTABLISH AREA OR TIME RESTRICTIONS?

The importance of clearly identifying the objective(s) for an area or time restriction in any particular instance cannot be over-emphasised (Chapters 1 and 5); unless a clear rationale for the action is developed it will be difficult to make appropriate decisions about how to implement the measure or to communicate and negotiate effectively with the relevant interested parties. As expressed in the definitions of an MPA noted above, a fishery manager might choose to adopt an area or time closure for a variety of reasons. The rationale underlying each of the objectives one might set are described below. Possible objectives have been classified here into three broad categories, the first dealing exclusively with fisheries management issues, the second with broader conservation considerations and the third with equity issues. In reality, the boundary between these categories is rarely clear cut, but it is convenient to make the distinction at this point.

2.1 As a fishery management measure

Paragraph 6.3 of the Code of Conduct articulates the general principle that "states should prevent overfishing and excess fishing capacity and should implement management measures to ensure that fishing effort is commensurate with the productive capacity of the fishery resources and their sustainable utilization". Area and time restrictions can assist fishery managers in achieving these objectives in the following ways.

2.1.1 Limiting harvest of specific life stages

Often it is desirable to prevent fishing on particular stages of a species life-cycle that are especially vulnerable to capture, or are critical to overall production. One example is of species that aggregate in particular areas to spawn; if fishing is allowed on these spawning grounds, it might not only disrupt reproductive activity in that year, but also unduly deplete individuals of reproductive age, leaving too few to contribute in subsequent years. If there are particular characteristics of the spawning habitat that might be affected by fishing on them, a permanent area closure may be required. Alternatively, closure of the area during the spawning season may be sufficient.

There might also be a need to protect areas where juveniles are particularly abundant; if a particular area contains a high proportion of juvenile fish along with adults, allowing a fishery to exploit the adults might lead to undesirably high levels of mortality on juveniles.

While protecting particular life-stages may require continuous closure of an area to fishing, it is often possible to only restrict fishing access during a particular season. The most appropriate measure will depend on the life-history characteristics of the species concerned, with closed seasons often being used for fast growing species with a short recruitment period, such as prawns and shrimps. In fisheries for such species, closing the fishery early in the season allows individuals to grow to larger and more valuable sizes.

2.1.2 Protecting depleted stocks and their habitats during the rebuilding phase of a fishery

If a fishery has collapsed, or is close to collapse, the action that needs to be taken to allow the stock to rebuild is likely to be draconian but essential (Code of Conduct, Paragraph 7.2.2). One option of course is to impose a complete ban on fishing. In some circumstances, however, it may be possible to protect stocks effectively with less stringent measures that allow fishing in some areas but prevent it in those that are critical to the rebuilding process.

2.1.3 Protecting genetic reservoirs

The value of a resource population being genetically diverse is important to appreciate, even though the benefits are often difficult to quantify. Understanding a few basic ideas helps one to appreciate why genetic diversity is important. The first of these is that the mortality fishing imposes often leads to differences in survival for fish with different characteristics. For example, most fishing is size-selective and removes large fish while leaving smaller ones. Fish in the population that start reproducing at small size are, therefore, likely to contribute more offspring to the next generation than those that wait until they are large, because the larger ones are more likely to be caught before they can reproduce. Thus, if the biological characteristic of size at first reproduction is passed on from parent to offspring (i.e. it is inherited), the inevitable consequence of size selective fishing will be that, over time, the average size at first reproduction of individuals in the population will fall. This is the process of natural selection. What does this mean for the fishery? In essence it means that adult fish will generally be smaller, which will usually be undesirable for the fishery. Maintaining a reserve in which larger bodied adults can persist may act as a genetic reservoir to counteract this trend.

Another important idea is that genetic variation provides insurance against changing environmental conditions. For example, some individuals in the population may grow better in warmer years and others when it is cold. If fishing reduces the population to very low levels it is possible that individuals with a genetic trait that may be important for the population in the future will be lost and the capacity for the species to adapt to a new situation compromised. Establishing protected areas to help preserve genetically diverse sub-populations may in some circumstances provide insurance against such possibilities.

2.1.4 Protecting habitat that is critical for the sustainability of harvested resources

Some types of fishing gear can have very large negative effects on benthic habitat that may be important for the sustainability of harvested resources. Often such habitats will be in inshore areas, where juvenile fish often aggregate in areas with high physical structure such as seagrass beds or mangroves. Among other things, the fish are afforded protection from predators in these areas. Paragraph 6.8 of the Code of Conduct makes specific reference to the importance of protecting such critical fisheries habitat as a guiding principle for responsible fisheries.

Although such habitats are more easily identified in shallow water, there may also be environments in deeper water that are important for similar reasons. In particular, structured habitat in deep water may provide refugia for commercially important juvenile fish. Structured benthic habitats are particularly at risk from mobile fishing gears, such as trawls and dredges (see Chapter 2), which can destroy them with only a few passes of the gear. Thus, it may be desirable to prevent access to trawlers and dredgers in such areas, while allowing, for example, pot or trap fishing.

2.1.5 To restrain excess fleet capacity and optimise the value of the catch

The Code of Conduct states that management measures should, among other things, provide that 'excess fishing capacity is avoided and exploitation of the stocks remains economically viable' (Paragraph 7.2.2a). When there is excess fishing capacity, a short appropriately chosen fishing season can optimise the value of the harvest while preventing over-exploitation of the stocks. Although by no means ideal from an economic perspective, in some cases this can lead to seasons that are restricted to a few days, when so-called "fishing derbys" or "the race for fish" takes place. In such cases, consideration needs to be given to how best to optimise the timing of the season. In the Bering Sea pollock fishery, for example, the opening of the season is delayed until late January when the pollock roe commands the highest price at market.

2.2 As a wider conservation measure

Coastal waters in particular are often rich in habitats that are highly valued for their aesthetic or other nature conservation values and some forms of fishing activity will alter such habitats in ways that harm those values. Permanent area closures in particular provide a mechanism for protecting such habitats, and their establishment is understandably a key goal for many sectors of the marine conservation movement. Importantly, MPAs for wider conservation purposes will normally seek to limit other activities in addition to fishing. Often, however, fishing is a primary target for restriction, partly because it is a practical proposition to impose restrictions, but also because, by definition, it directly exploits a biological resource. Other less direct impacts, such as pollution inputs from diffuse sources on land, are more difficult to deal with, especially given the open nature of marine systems and high rates of exchange.

An indication of the value and importance of the MPA as a conservation measure is provided by the fact that the IUCN and others have called upon national and inter-governmental agencies to adopt a series of goals centred upon them. Specifically, they have argued for a global representative system of marine protected areas in accordance with a set of guiding objectives (Table 1) and for national governments to also set up their own systems of marine protected areas; a number of nations have already taken such steps, including Australia, Canada and the USA, with other nations likely to follow suit in the future.

2.2.1 Protecting benthic habitats of high conservation value

As noted above, structural epibenthic communities can be especially vulnerable to towed fishing gears. When such areas have been identified to be of high conservation value, establishing a permanent closed area that prevents fishing by such methods is probably the only measure that will protect them. Such protection is endorsed by the Code of Conduct under Paragraph 7.2.2d, which states that management measures should provide that 'biodiversity of aquatic habitats and ecosystems is conserved and endangered species are protected'.

2.2.2 Limiting bycatch

In some groundfish fisheries, for example off Alaska, closed seasons have been set to minimise bycatch rates or potential interactions with marine mammals.

Table 1: Guiding objectives for the establishment of a representative system of marine protected areas. Extract from the Resolution of the 4th World Wilderness Congress, Colorado, USA, September 1987.

- To protect and manage substantial examples of marine and estuarine systems to ensure their long-term viability and to maintain genetic diversity
- To protect depleted, threatened or endangered species and populations and in particular to preserve habitats considered critical for the survival of such species
- To protect and manage areas of significance to the life cycle of economically important species
- To protect outside activities from detrimentally affecting the Marine Protected Areas
- To provide for the continued welfare of people affected by the creation of marine protected areas; to preserve, protect, and manage natural aesthetic values of marine estuarine areas, and historical and cultural sites for present and future generations.
- To facilitate the interpretation of marine and estuarine systems for the purposes of conservation, education, and tourism.
- To accommodate within appropriate management regimes a broad spectrum of human activities compatible with the primary goal in marine and estuarine settings.
- To provide for research, training and monitoring of the environmental effects of human activities, including the direct and indirect effects of development and adjacent land use practices.

2.2.3 Protecting attributes of the ecosystem that are critical for maintaining ecosystem services

The idea that ecosystems provide services to mankind is one that has only recently emerged (Costanza et al., 1997). In essence, the term 'ecosystem service' connotes the idea that ecosystems serve functions that are of value to mankind and that maintaining the basic structure of the system will ensure that these functions continue to operate. By preventing fishing activity in particular areas, it is argued that ecosystem services will be preserved. These services can be divided into two classes – extractive services (things that provide products, etc.) and existence services (things that you get, just because something exists). Of course, extractive ecosystem services include the provision of fish, oil and mineral resources, the management of which may be served in part by access restrictions to particular areas (see above with respect to fisheries management). Of more relevance to this discussion, however, are the existence services, examples of which include water purification and nutrient regeneration. Unfortunately, for the fisheries manager, while the need to protect ecosystem services is a point that may often be raised in consultation with other interested parties, adequate operational frameworks for defining such services and the protection that is needed will often be lacking.

Despite the above difficulties, however, certain water bodies (e.g. reed beds, wetlands, mangroves or lagoonal areas) might, for example, be shown to be important for protecting coastal environments by removing high nutrient loads from land run-off before entering the sea. In this case, protecting such habitats from fishing activities such as trawling might be justifiably argued for on the basis that it would preserve an important ecosystem function.

Notwithstanding the comments above, a manager should be aware that other justifications for permanent area closures, based on arguments about ecosystem function, may be much more difficult to make. It is often claimed, for example, that high levels of biodiversity are important for ecosystem function. While there are many very valid justifications for protecting biodiversity, the evidence for a positive relation between this attribute of the ecological community and ecosystem function remains a subject of considerable scientific debate. A manager should be wary, therefore, about giving undue weight to justifications for establishing an MPA on biodiversity-function grounds. There may well be other factors that are far more important for preserving function than preserving the number of species an area supports and other much sounder arguments for establishing an MPA that do not depend on such contentious scientific hypotheses.

2.3 To resolve equity issues

2.3.1 Providing a mechanism for resolving conflict over multiple-use of areas or resources

The coastal zone in particular is an area where a multiplicity of users require access. Often, however, uses are incompatible with one another. Trawling in an area where pot fishermen operate, for example, can lead to major conflicts when pots are destroyed. Similarly, combining a submarine practice area with trawling activity would be a bad idea! There are many other possible conflicts of use (e.g. tourism, shipping, recreational fishing), where the only tractable solution is to restrict activities to certain areas by some form of zoning arrangement, either on a permanent or seasonal basis.

2.3.2 Reserving economically vital marine and coastal resources for the preferred use of residents or traditional users

Often indigenous cultures have traditional (and sometimes exclusive) claims on certain lands or resources that can be served by the establishment of some form of area closure or exclusive season. Similarly, local fishermen's cooperatives or communities might benefit from rights protection that is area based. In some cases the fishery manager will also need to incorporate issues of non-governmental management and customary tenure into the equation. The Code of Conduct (Paragraph 6.18) calls on States to "appropriately protect the rights of fishers and fisherworkers, particularly those engaged in subsistence, small-scale and artisanal fisheries..."

3. WHAT ARE THE ADVANTAGES AND DISADVANTAGES OF AREA AND TIME RESTRICTIONS?

3.1 Advantages

3.1.1 Conceptual simplicity

There is no doubt that it is relatively easy to explain and justify to many sectors of the community the reasons for establishing, and the mechanisms for implementing, area or time closures. To use a terrestrial analogy, placing a fence around a piece of land clearly establishes a property right and identifies the basis for its use. Dividing up the fishing season, or fishing grounds, among different fishing sectors is an obvious solution for resolving access issues. In principle at least, it is also a relatively straightforward matter to specify an area or time closure in legislative terms once agreement has been reached among interested parties that the measure is appropriate. With respect to enforcement there are also clear advantages to a reserve system or fishing season when the local fishing community supports the initiative and polices it themselves. For example, control of fishing effort by closed areas or seasons seems to be one of the few options open to managers of marine municipal fisheries in the Philippines (see below).

3.1.2 A good option for protecting bycatch species that cannot be protected by other means

When bycatch species are at serious risk, closed areas or seasons offer a means for protecting them. It is important to recognise, however, that the appropriateness of such approaches critically depends on a clear understanding of the life-history of the species concerned. For example, for highly mobile species that range over an entire region a closed area may be completely ineffective or need to be impracticably large.

3.1.3 A tractable approach for stock protection in complex fisheries or when data are poor or absent

The potential for permanent area closures to be a cost-effective means for managing fisheries for coral reef systems has been recognised for some time, but it is also now being widely advocated for some temperate fisheries. In reef systems, the major objective for such reserves is to protect critical spawning stock biomass to ensure recruitment supply to fished areas via larval dispersal, and possibly to maintain or enhance yields in areas adjacent to reserves. If they work, they have the additional advantage of being easier to implement than more conventional fisheries management programs. This aspect makes them especially attractive for coral reef fisheries which are almost always multi-species, with many artisanal or subsistence fishers using a wide variety of gears and landing their catch at many sites over a wide area. These features make it difficult to collect even the most basic information such as catch and effort that are required for conventional management.

3.1.4 A sound approach for protecting sensitive benthic habitats

It is axiomatic that if one wishes to protect sensitive sessile benthic communities, which often have recovery times of 10-15 years following significant impact, some form of area closure will often be the only alternative when fishing activity has been shown to be a threat.

3.1.5 Insurance against uncertainty

A fishery manager is often faced with the problem that the information which scientists can gather and the predictions they are able to make are uncertain. He or she must also be mindful of the need to adopt a precautionary approach as articulated by the FAO in its Technical Paper on the precautionary approach to fisheries (FAO, 1996). Indeed, much of the focus of fisheries science is currently on finding ways to quantify and communicate that uncertainty to managers so that they can make informed decisions (see Chapter 5). Along with uncertainty comes the need for insurance policies that provide some degree of safeguard in the event that decisions are based on overly optimistic predictions.

Closed areas can sometimes be viewed as providing such insurance. For example, a reasonable management goal for demersal fish resources might be that the stock should remain at above 60% of the un-exploited biomass over a given time horizon, say 20 years. Maintaining such levels would put the stock in the region of optimal sustainable yield that many fisheries aspire to. Using this goal, recent theoretical analyses for a large-scale demersal fishery on the continental shelf suggest that establishing an MPA could be an important bet-hedging strategy and could act as an effective insurance policy that would protect both the long-term future of stocks <u>and</u> yield higher average catches (Lauck et al., 1998). It should be stressed, however, that in this theoretical exercise, the MPA had to be very large to be effective. This latter conclusion is supported by independent work undertaken by the International Council for the Exploration of the Sea (ICES) to examine the utility of a closed area in the North Sea to protect cod stocks. Given current understanding of fish movements and the behaviour of fishing fleets, even closing one quarter of the North Sea to fishing would do little or nothing to protect the widely dispersed and mobile cod. It should also be stressed, therefore, that although MPAs can theoretically serve to protect stocks, it is by no means axiomatic that they will do so in all cases (see Section 4 of this Chapter).

3.1.6 A tool for continuous improvement

An important benefit from establishing an area or season closure is that it can provide an area for research to learn more about how a marine system works, or it can be viewed as a form of management experiment to gain information that will lead to better decision-making in the longer term. One could view such benefits as a primary objective for an area or time closure, but more often it will be an ancillary benefit that flows from the establishment of the area for other reasons. Such approaches can sometimes fall in the realm of what has been termed adaptive management, where new measures are tried specifically to test ideas and learn more so that management can be adapted in the light of results. A good example of this approach is provided in Section 4.

3.2 Disadvantages

3.2.1 Inter- Agency negotiation

When the objectives for establishing an area or time closure require restrictions on activities other than fishing, it will be necessary to negotiate with other agencies and interested parties. This can often be an extremely long and drawn-out process that requires considerable negotiating skill and political judgement. Critical to overcoming the difficulties inherent in the process is a clear and agreed set of shared objectives for the measure.

3.2.2 High economic cost in some cases

Although an area or time closure may be effective in conserving a stock, other measures may be more desirable on economic grounds. Displacing fishing effort from areas or times that are economically optimal can be very costly in terms of the economics of the fleet.

3.2.3 Reduced effectiveness of restrictions over time in the absence of complementary measures

In cases where a time or area closure is imposed to limit catches by effectively reducing effort, then without limits on capacity, effort or catches, the effect is to encourage an increase in capacity over time that eventually undermines any short-term benefits of reduced catches. Such responses serve to emphasise that area or time closures will usually need to be combined with other input and output controls to ensure that the stocks are well managed.

3.2.4 Enforcement

Although establishing an area or time restriction might look good on paper, without a convincing enforcement mechanism the measure is clearly pointless (Chapter 8). As noted above, when fishers support the measure there can be a strong incentive for self-policing, which can make enforcement relatively straightforward. In other circumstances, however, particularly when a fishery operates over a large or remote area, the practicalities and costs of enforcement can be prohibitive. It seems likely, however, that in some cases satellite surveillance techniques could be used to good effect – (Section 3.2.6, Chapter 8).

3.2.5 Enthusiasm versus Appropriateness

There is considerable enthusiasm for the establishment of permanently protected areas by many sectors of community. Unfortunately, however, this enthusiasm may lead to un-sound judgement concerning the likely effectiveness of such a measure for achieving specific objectives. This possibility is particularly likely with respect to fishery management, where considerable uncertainties often remain concerning the effectiveness of permanent "no take" zones. Careful quantitative evaluation of the benefits that will flow from the establishment of an MPA is highly desirable when the objective is stock protection. It may well be, for example, that using other input or output controls is more appropriate for managing any particular fishery. For wider conservation purposes, however, the establishment of a permanent area closure may better serve objectives. Indeed, if one wishes to protect sensitive habitats from mobile fishing gears it is difficult to see an alternative. This example serves, once again, to emphasise that clarity of objectives is of paramount importance.

4. CASE STUDIES

4.1 Gulf of Mexico: a mixture of area and time closures

Good examples of seasonal closures can be found in many fisheries, but shrimp fisheries seem especially suited to the approach because juveniles generally develop in coastal estuarine environments and move offshore as they continue their life cycle, which is completed often within about a year. Thus, to prevent growth overfishing, where individuals are harvested at sub-optimal size, the fishery is closed during the early period in the growing season. Authorities in Texas for example, close state and federal waters from mid May to mid July – a measure that protects juvenile shrimp migrating from the bays to the Gulf of Mexico, thereby allowing them to grow to a larger more valuable size.

A good example of a combination of area and time closure to minimise conflict between fishing sectors can also be found in the Gulf of Mexico, where the State of Florida has a closure zone which restricts the trawl fishery for shrimp and trap fishery for stone crab (Figure 1).

4.2 Coral reefs

There are a number of practical examples of the closed area approach in coral reef systems, with a number of authorities imposing temporary or permanent closure to fishing of portions of or, in some cases, whole reefs. Often, this has been done in the hope that it will prevent fish stocks being depleted and maintain or even enhance yields in adjacent areas. Additional benefits might accrue if tourist activity is to be promoted in an area. Unspoilt coral and an abundant fish life are a pre-requisite for attracting visitors. Of course, given the destructive nature of some forms of fishing and tourism activity (especially poisoning and dynamite fishing, which although illegal, still occur) the imposition of closed areas on coral reefs serves a conservation purpose that does not require positive benefits to the fishery. Obtaining these benefits may be justification enough in some cases with no need for additional benefit to tourists or fishers.

In the case of coral reefs, studies have shown that after affording protection, even to relatively small areas, the densities and biomasses of target species generally increase within the reserve area (see Hall, 1999, for review). Perhaps the best demonstration of this effect is from the Philippines, where fish populations were compared in two small areas (Sumilon and Apo) where protection from fishing was variously established and then relaxed over a period of 10 years. Figure 2 summarises the results of this study. For Sumilon densities of large predatory fish decreased significantly when it was opened to fishing in 1985 and 1993, and increased significantly three times following periods of protection. In contrast, at Apo there was a steady increase in densities over an eleven-year period of protection while comparable non-protected areas showed little change.

The above results and other similar results from other areas show that fish populations in both temperate and tropical regions respond, even in relatively small areas, if one protects them from fishing. In Kenya, for example, studies on two reefs - one in a marine park that allowed no fishing and another in a reserve that only allowed artisanal fishing – found that the abundance of commercially important species was ten times greater in the fully protected area. One rather unsettling observation from the Philippines study described above, however, is that it took only 1.5 and 2 years of unregulated access to an area to eliminate density and biomass gains accumulated over 5 and 9 years of marine reserve protection.

Another response that is observed is an increase in the number of species found in protected areas. For example, studies in Kenya have shown that 52 of 110 species that were found on protected Kenyan coral reefs were not found on un-protected reefs – of these, 44 species were unique to the protected coral reefs. Similar responses have been observed in a number of other studies; if the objective is conservation, marine reserves would seem to be very effective.

Although increases in abundance (and perhaps of species richness) within reserves is a general rule, the establishment of an MPA for fisheries management purposes will generally need to be justified on the grounds that the higher spawning stock biomass inside the reserve will contribute recruits to the adjacent fishery. Since adult fish that inhabit reef systems tend to be fairly immobile this is often a realistic proposition.

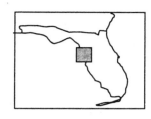

- No person shall operate any trawl in the following zones during the periods indicated:

a) Zones I and III: Oct 5th-May 20th of the following year
b) Zone IV: Dec 2nd-Apr 1st the following year
c) Zone V: Dec 2nd –Mar 15th the following year

- No person shall fish with, set, or place any stone crab trap in the following zones during the periods indicated:

a) Zone II: Oct 5th – May 20th of the following year.
b) Zones IV & V: Oct 5th – Nov 30th of the same year and Mar 16th – May 20th of the same year.

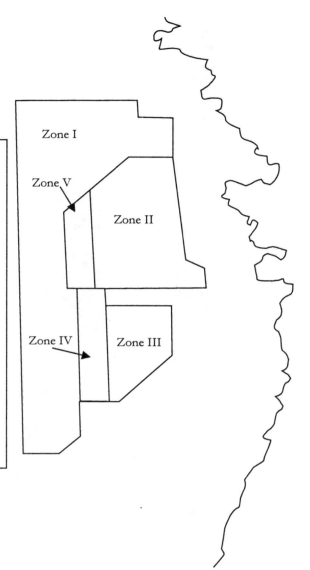

Figure1: The State of Florida Citrus-Hernando Shrimp/Stone Crab closure zone, a 144,000 acre closure area, which restricts access by each fishery to designated areas of seabed for defined periods each year.

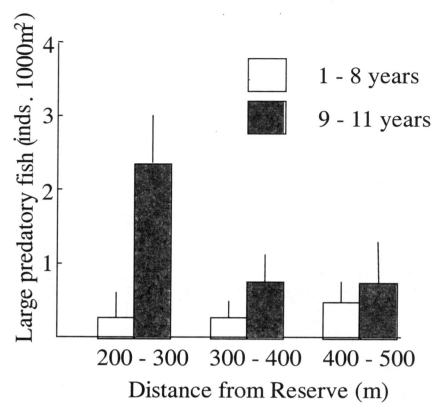

Figure 2: The densities of large predatory fish at different distances from the Apo reserve boundary during the first 8 years of reserve protection and from years 9 to 11. Data are means ±1 standard error of the mean. Redrawn from Fig. 3 of Russ & Alcala (1996) and reproduced from Fig. 9.1 in Hall (1999).

4.3 Surf zone fisheries

One other tropical example, where there is good evidence that a marine reserve increases yields in adjacent areas fished comes from the De Hoop Marine Reserve on the southern African coast (Attwood and Bennett, 1994). Here, tagging studies over a five-year period showed that Galjoen (*Coracinus capensis*), a species exploited by anglers, showed two distinct behaviours. Part of the population was relatively sedentary with home ranges within the reserve, while the other part was nomadic. Estimates of the numbers dispersing strongly suggest that the reserve, which spans a 50 km stretch of coastline, was contributing to the fishery by providing a supply of mature fish to both nearby and distant exploited areas.

4.4 Georges Bank

On Georges Bank, off the northeast coast of the United States, as with many other temperate demersal systems, controls on mesh sizes, minimum fish sizes and seasonal areal closures failed to conserve stocks because there was no direct control on fishing effort. Changes in fish community structure occurred largely as a consequence of highly species-specific harvesting patterns driven by market considerations. In response, the authorities set up long-term areal closures in 1994 to try and improve the fisheries of the region (Fogarty & Murawski, 1998). These areas encompass areas of traditionally high catch-per-unit-effort (i.e. good fishing grounds), including part of the scallop grounds of the region and important spawning grounds for cod, haddock and yellowtail flounder. Sand/gravel areas that may be important for, among other things, juvenile survivorship are also protected. The effect of this closed area was described recently by Murawski *et al* (2000), who showed that, although whitefish stocks have increased during the period of closure, other management measures had also changed over the period. This has made determining the cause of the increase difficult. For scallops, however, the size of stock increased considerably, which was almost certainly in response to the closure.

4.5 The Plaice Box

The Plaice Box is an area of about 38,000 km^2, along the Danish, German and Dutch coasts that was established in 1989 to protect juvenile flatfish (plaice and sole) by preventing large vessels from fishing in the region in the second and third quarter of the year. In 1994 an analysis of the plaice box was undertaken, which explored the benefits of various management options by comparing expected long-term landings and spawning stock biomass (SSB) of plaice and sole, compared to the *status quo*. This analysis indicated that, if the box were removed, long-term landings and SSB would decline by 8-9%, but if the prohibition were to be extended to the entire year, landings and stock biomass would increase by 24-29% (Table 2). The main reason for these substantial benefits stemmed from the fact that previous discarding rates inside the box averaged 83%. In the light of these analyses, new regulations were established which extended restrictions to all year, but which permitted selected vessels to continue fishing, particularly those targeting shrimps.

Table 2: Percentage increase in long-term landings and spawning stock biomass (SSB) for the North Sea fleet relative to the 1994 *status quo* quarter 2 and quarter 3 closure of the Plaice Box (From Horwood, 2000).

Management option	Landings (%)	SSB (%)
Remove Box	-8	-9
Extend to quarter 4	+11	+14
Extend to entire year	+14	+17
All year + no discarding fleets	+24	+29

4.6 Adaptive management using the closed area approach: an example for NW Australia

As noted in Section 3.1.6, one of the advantages of implementing closed areas is that they provide an opportunity to learn more about the marine environment and progressively improve management decisions. Perhaps the best example of such an approach comes from the northwest shelf of Australia, which also illustrates the importance of including the possible effects of fishing on fish habitat into decision-making.

Research survey data available from 1960 onward showed that, while the total biomass of fish had not changed as the fisheries of the region had developed, the composition of the fish community had altered, with *Lethrinids* and *Lutjanids* declining and *Saurids* and *Nemipterids* increasing (Fig. 3).

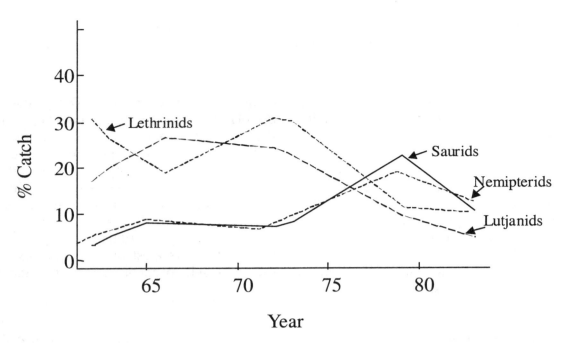

Figure. 3: Trends in abundance of the four major exploited fish taxa on the Australian Northwest shelf. Adapted from Fig 14.2 in Sainsbury (1988) and reproduced from Fig. 3.7 in Hall (1999).

The available data also indicated that the benthic environment had altered over the same period. In particular, the quantity of epibenthic fauna caught in trawls (mainly sponges, alcyonarians and gorgonians) is now considerably lower than it was prior to the development of the trawl fishery (Sainsbury, 1987). Underwater video data indicated four habitat types in the region on the basis of dominant benthic fauna, and fish catch data indicated that *Lethrinids* and *Lutjanids* were almost exclusively associated with habitats supporting large epibenthos. In contrast, the lower value *Saurids* and *Nemipterids* were only found on open sand.

This information gave rise to important management questions for the region: 1. Could the change in fish and benthic community composition be reversed? 2. If changes were reversible, was it worth attempting to do so given the uncertainties of the outcome and the time frame over which the change would occur? 3. If an attempt was to be made, what management measures were most appropriate to achieve the goal?

The key to resolving these issues lies with understanding the mechanisms responsible for the changes in the first place; four alternative hypotheses were formulated:

1. *intra-specific dynamics*: the observed changes result from independent responses of each species;

2. *competitive release due to fishing*: there is a negative influence of *Lethrinus* and *Lutjanus* on the population growth rate of *Saurida* and *Nemipterus* so that when the *Lethrinids* and *Lutjanids* were removed by fishing the latter experienced a release from competition and increased in abundance;

3. *competitive depression*: *Saurida* and *Nemipterus* have a negative influence on the population growth rate of *Lethrinids* and *Lutjanids* and the abundance of these species declined because the former increased for reasons independent of the fishery;

4. *habitat modification*: habitat characteristics determine the carrying capacity of each genus separately so that trawl-induced modification of the abundance of the habitat type alters the carrying capacity of the different genera.

All four hypotheses are ecologically reasonable and were consistent with the available data. It is important to recognize, however, that each has markedly different management implications. Hypotheses 1 and 2 imply a relatively low productivity of *Lethrinids* and *Lutjanids* with reductions in the biomass of these taxa being viewed as a consequence of fishing. Accordingly, even if stocks could be rebuilt, the sustainable yield from the fishery would need to be low to prevent the same decline happening again. In contrast, hypotheses 3 and 4 imply a relatively high productivity for *Lethrinids* and *Lutjanids* under some circumstances. Selective harvesting of these taxa under hypothesis 3 and harvesting without damage to benthic habitat structure under hypothesis 4 would result in comparatively high sustainable catches. These differing implications make determining which mechanism operates much more than an academic exercise.

To address these issues, a formal evaluation procedure was initiated by stating the above hypotheses as explicit mathematical models. Establishing such models is very worthwhile because they permit one to formally evaluate which hypothesis is most likely to be true given the available data.

The statistical analysis suggested that there was a relatively low expected present value from continuing the existing licensed trawl fishery and the additional information that could be gained from monitoring the outcome of continued trawling would not help make future decisions about what to do for the best. In fact, even though the probabilities that could be assigned to the various models at the time were relatively low, there appeared to be clear benefits from an immediate switch to a domestic trap fishery. However, it was also shown that some experimental management regimes, involving cessation of trawling in some areas and the introduction of trap fishing in some of the areas closed to trawling, could provide a higher expected present return from the resource.

Partly on the basis of this work, the management agencies for the northwest shelf agreed to conduct an experiment by subdividing the area into three zones. One part of the area was left open to trawlers, a second part was closed to trawlers in 1985 and a third was closed in 1987. Trap fishing was permitted throughout. It was hoped that closing part of the area to trawls would allow this fishery to develop to exploit species which are found in less disturbed habitats.

Despite some difficulties, an adapted experiment is still continuing and sufficient data have been collected to allow the four hypotheses to be evaluated (Sainsbury et al., 1997). Figure 4 shows how the area closed to trawling experienced an increase in the density of *Lethrinus* and *Lutjanus* and in the abundance of small benthos. The abundance of larger epibenthos stayed the same or perhaps increased slightly. In the area open to trawling, the abundance of fish declined along with the small and large epibenthos.

These results provide a valuable perspective on fishery effects, but it is in the formal evaluation of the four mechanistic hypotheses mentioned above that the real strength of this study lies. This is because the data from the experimental period allowed the probabilities assigned to each of the four hypotheses to be further updated. These updated results indicate that a high value *Lethrinus* and *Lutjanus* fishery could be established on the northwest shelf if the habitat could be protected and that changes in fish community structure can probably be attributed in large part to habitat modification by trawling. Managers who are able to draw on such quantitative and systematic scientific evaluations will be in a much better position to make informed decisions.

The northwest shelf is a good example of where an interaction between fisheries and the structure of benthic communities may lead to both an enhanced fishery and a less disturbed benthic community. Such mechanisms may not happen everywhere, indeed, the habitats in which they operate might be quite restricted, but one should be alert to the possibility. Such effects need to be considered when determining suitable technical measures to use in a management strategy (Chapter 2).

Unfortunately, however, in the case of the northwest shelf, it is apparent that the time scales for recovery for epifaunal benthos are slower than previously thought. Rather than taking 6-10 years for sponges to grow to 25 cm, it now appears that at least 15 years are required. Moreover, video analysis of the effects of the trawl ground rope indicated that about 89% of encounters lead to dislodgement of sponges and almost certainly subsequent death. This slow recovery dynamic and the apparently high probability that large benthos will be removed by a trawl mean that measures to protect the habitat would need to be very effective to maintain the habitat structure required to support this high value fishery.

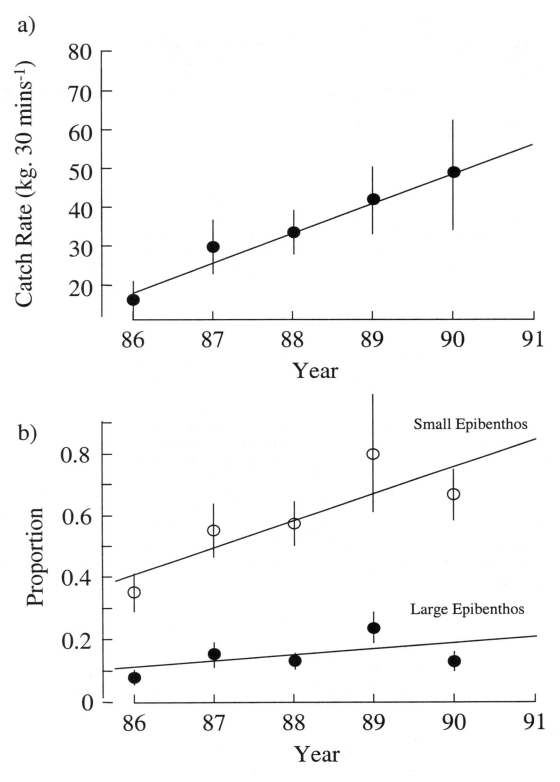

Figure: 4. a) Changes in the abundance of fish (*Lethrinus* and *Lutjanus*) b) Changes in the proportion of large and small epibenthos in areas closed to fishing. Adapted from Fig. 2 of Sainsbury et al. (1997) and reproduced from Fig. 3.8 in Hall (1999).

5. WHAT ARE THE PRACTICAL STEPS TOWARDS ESTABLISHING TIME AND AREA RESTRICTIONS?

A brief summary of the steps involved in the implementation process, for fisheries management or conservation purposes is given in Figure 5. Further commentary on aspects of the process are given below.

5.1 Set your goal

The importance of being explicit about the goals for area or time restrictions has been continually emphasised throughout this chapter and is throughout this Guidebook. To re-iterate, it is essential that the manager selects from among the justifications presented in Section 2 or identifies an alternative goal. There may, of course, be multiple justifications for the measure, in which case it is important to try and specify them in order of priority. While there can be no hard and fast rules about how detailed the specification of objectives for a measure should be, its establishment will be greatly facilitated by including as much detail as possible about what you seek to achieve at the earliest opportunity. This will require translating the broad goals into detailed operational objectives (Chapter 5).

5.2 Specify criteria for selection

5.2.1 Criteria for stock management

Once the objectives for establishing an area or time restriction have been established, the criteria for the selection of candidate sites or time periods should follow logically. For many time and area restrictions, the choice will often be driven by the life-history of the species involved and the dynamics of the various fishing sectors that the measure is designed to serve. Adequate biological data to support the decision will often be available - the location of nursery grounds and spawning seasons or areas are often relatively well known. Similarly, fishers are acutely aware of where and when conflicts between sectors occur. With respect to the justification of permanent no-take zones in the adult habitat, however, data will often be less comprehensive and the potential benefits more open to debate. Lauck et al. (1998) note the following as "desirable" features for a no-take reserve, established to fulfil fisheries management objectives.

1. It should be large enough to protect the resource in the event of overfishing in the unprotected area.

2. The reserve should serve as a source capable of replenishing the exploited stock in the event of its depletion. In particular, reserves should protect spawning grounds and any other areas critical to the viability of the population.

3. The reserve must be completely protected, since the almost certain build up of biomass inside the reserve will be very attractive to poachers.

In my view these features are not just desirable, but essential if an MPA is going to work to protect stocks.

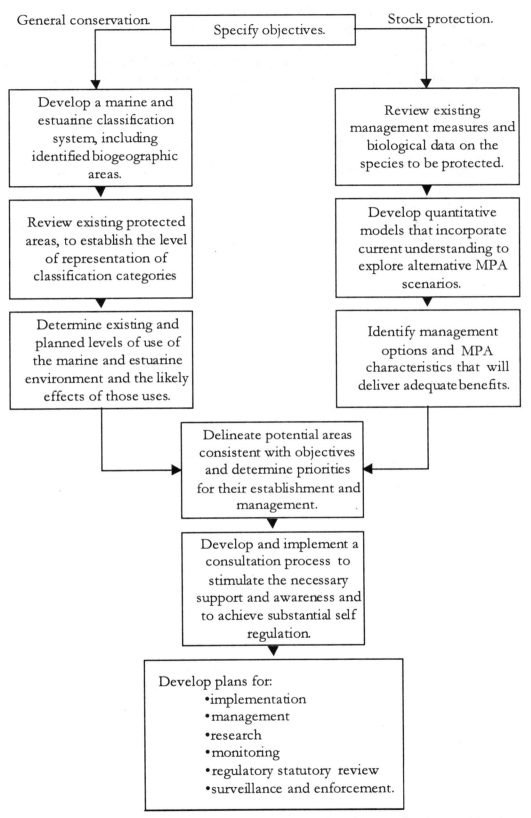

Figure 5: A diagrammatic summary of the steps involved in the specification and implementation of a MPA for either fisheries management or other conservation purposes. Adapted in part from Kelleher & Kenchington (1992).

With respect to point 3, there should be relatively little difficulty in working out what is required – it is simply a matter of establishing the political will to meet the costs. In contrast, points 1 and 2 are a major challenge for ecologists. Identifying appropriate sizes and locations for closed areas requires consideration of the relative proportions of the populations and communities of interest within the protected region, their potential to serve as source populations for unprotected areas, and the location of any sensitive habitat types, which need to be included in the protected area to maximise the benefits of the approach.

The preferred model for area and time restrictions for more general conservation purposes is for legislation that is based upon sustainable multiple-use managed areas. Isolated highly protected pockets in an area that is otherwise unmanaged or subject to piecemeal regulation is much less desirable because piecemeal protection of small marine areas alongside conventional fisheries management often leads to over-exploitation of fish stocks and progressive deterioration of the protected area. Thus, most conservationists favour larger multiple-use protected areas that provide for a number of levels of access and of fishing and collecting in different zones, with continued sustainable harvest of food materials in the majority of a country's marine area. A summary of the IUCN criteria for the inclusion of an area as an MPA are given in Table 3.

Table 3: The factors or criteria that have been proposed for deciding whether an area should be included in an MPA or for determining the boundaries of an MPA. Adapted from Kelleher & Kenchington (1992).

Criteria	Description
Naturalness	The extent to which the area has been protected from, or has not been subject to, human induced change.
Biogeographic importance	Either contains rare biogeographic qualities or is representative of a biogeographic "type" or types. Contains unique or unusual geological features.
Ecological importance	Contributes to maintenance of essential ecological processes or life support systems. The degree to which the area either by itself or in association with other protected areas encompasses a complete ecosystem. Contains some or all of the following: • a variety of habitats; • habitat for rare or endangered species; • nursery or juvenile areas; • feeding, breeding or rest areas; • rare or unique habitat for any species; • high genetic diversity i.e. diverse or abundant in species terms.
Economic importance	Existing or potential contribution to economic value by virtue of its protection. Economic contribution could be delivered through: • recreation; • subsistence; • used by traditional users; • appreciation of tourists; • important habitat for economic importance species.
Social importance	Has existing or potential value to the local, national or international community owing to its heritage, historical, cultural, traditional, aesthetic, educational or recreational qualities.
Scientific importance	Value for research and monitoring
International or national significance	Has the potential to be listed on the World or National Heritage list or declared as a Biosphere reserve, of other national or international significance, or subject of a national or international conservation agreement.
Practicality/feasibility	• Degree of isolation from external destructive influences • Social and political acceptability, degree of community support • Accessibility for education, tourism, recreation • Compatibility with existing uses, particularly by locals • Ease of management, compatibility with existing management regimes.

5.3 Assemble information and conduct a preliminary evaluation

Clearly, to address the criteria described in the previous section a considerable amount of economic, social, biological and ecological information is required before informed decisions can be made. As noted above for many of the objectives one might set for area and time restrictions, biological and fishery data will be adequate. For some forms, particularly of permanent area closures in adult fish habitat, the information for basing a decision about a closed area will, in many cases, be poor or unavailable. Whatever, the available data, a manager must seek to learn from past experience in other systems and, if possible, from scientifically defensible quantitative models that predict the likely consequences of different management scenarios (Chapter 5). Questions that it may be important to ask include the following.

5.3.1 Will my restriction protect fish stocks?

In the limit, of course, where the vast majority of an area is closed to fishing, the answer has to be yes (unless the small piece that was left open happened to be the only spawning ground for the resource). A more appropriate question, however, is under what range of circumstances is area or time restriction likely to succeed in this objective and by what mechanisms? This is a complex and difficult question, for which no simple answers can be offered, especially for permanent area closures in adult fish habitat. It is perhaps notable, for example, that the Organization for Economic Cooperation and Development (OECD) recently reviewed the benefits to fisheries of 52 restricted areas and found that in 32 cases stocks declined or showed major oscillations and in only 16 had stocks increased or remained the same (Anon 1997). Importantly, in all of the successful cases limited entry or TACs were also used along with other input controls such as size or sex selectivity so the contribution made by the area closures alone to the outcome was impossible to determine.

Despite this somewhat pessimistic OECD analysis, however, there is, good evidence that for many purposes area and time restrictions have been highly successful. Indeed, as noted above, there is also growing evidence for successful MPAs in adult habitat for fish stock conservation in coral reef and other tropical systems, where the biology of the fish is favourable and management is through effort control. One might expect similar positive benefits in temperate systems where the fish species of interest have similar life history characteristics to their tropical counterparts. At present, however, the case for permanent closures is rather less convincing for temperate continental shelf fisheries, although seasonal closures such as the Plaice Box described above certainly appear to confer benefits.

5.3.2 How big would my MPA have to be or how long should restrictions run for?

As noted earlier, for fishery management in temperate demersal systems, area closures in adult habitat may need to be especially (and perhaps unfeasibly) large to be effective. This may be especially true in fisheries that are managed through controls on total allowable catch, where effort may be diverted to other areas and negate potential benefits (Horwood, 2000). Horwood argues convincingly, for example, that claims for benefits to fish stocks using closures of the order of 10-20% of fishing grounds are overly optimistic for such systems in Europe. However, seasonal or other closures to protect juveniles certainly appear to be effective. With respect to coral reef systems, McClanahan & Kaunda-Arara (1996) suggest that many small reserves may be preferred from a fisheries perspective. These authors found that small reserves increased the total catch in adjacent areas, but a larger park did not. This effect may be due to the lower ratio of edge to park area in a large reserve. Data or experiences for tropical demersal trawl fisheries appear to be lacking and there is little analysis available to guide decision-making for these circumstances.

For more general marine conservation, the concept of a park or reserve in a marine system is somewhat different to that which obtains on land. In terrestrial systems one generally thinks of a protected area as being separate from the rest of the system, with surrounding unprotected areas having little influence within the park. In contrast, marine systems are usually extremely open with considerable exchange across the legislative boundaries defined by lines on a map. It follows that the minimum size required to meet the objectives for an MPA is often likely to be many times larger than that required for a terrestrial system. In particular, there is general agreement that, in order to protect productive marine ecosystems and areas of high biodiversity, an MPA needs to encompass as many of the ecosystem components as possible and give full consideration to the many factors and influences affecting productivity and diversity.

5.3.3 Who are the interested parties that will be impacted by the measures and what are the legislative issues associated with its implementation?

Decisions about closed seasons will usually fall exclusively within the domain of the fisheries manager, with little need to involve agencies or interested parties other than those with a direct interest in fishing. In contrast establishing permanent MPAs will almost certainly require inter-agency agreements and negotiations. An important goal for a fisheries manager will be to acquire the knowledge required to support the planning and management process. Indeed the general principles outlined in Paragraph 6 of the Code of Conduct make a number of references to the importance of consultation and negotiation. In particular, the implementation of an MPA is more likely to be successful if:

- the foundation for planning is based on real biological and physical data;
- this foundation includes data on cultural aspirations and the socio-economic position of interested parties;
- the facts are available in a readily understandable format to explain and, where necessary, justify concerns and actions;
- there is a clear consultative process established (Chapter 7).

With regard to legislation, it is clearly the case that management success is more likely when the area under consideration for protection (and preferably the adjacent land in coastal waters) is under the same agency's jurisdiction. In most cases, however, this is unlikely, and in cases where this situation does not obtain, it has been strongly argued that legislation and management arrangements should grow from existing institutions unless there is overwhelming public or political support for completely new administrative agencies. Of course, the precise legislative environment in which an MPA is to be established will differ from country to country. Thus, no general guidance can be provided here beyond emphasising the value to a manager of being fully acquainted with the legislative issues for their own situation.

5.3.4 How would the measure be enforced?

As with all fisheries management measures, the ease with which compliance can be monitored and enforced is likely to be a critically important determinant of feasibility. Unfortunately, for area and time restrictions, there is no straightforward answer to how easy enforcement will be. An analysis by the OECD (Anon, 1997), for example, showed that increased enforcement costs or problems were reported for six fisheries, while five fisheries reported no difficulties. One could imagine a situation, for example, where a short fishing season for boats from a limited number of ports could easily be policed. In contrast, permanent closure of a large and remote area would be almost impossible to control without technological support such as satellite tracking of vessels, or air surveillance. Without doubt seasonal or area closures are likely to be most effective where the fishers themselves agree wholeheartedly with the measure and are prepared to police it themselves (Chapters 7 and 8).

5.3.5 Would there be a need for adjustment funding?

An important consideration when access to fishing grounds is to be restricted is the extent to which effort is likely to be displaced elsewhere and what the consequences of such displacement are likely to be. If effort displacement is a concern, the possibility of providing funding for structural adjustments to reduce capacity and compensate fishers for loss of access may need to be considered.

5.4 Implement the process of negotiation

After undertaking a comprehensive evaluation of the options and becoming familiar with the scope of the problems for a given situation, the fisheries manager should be in a position to decide whether an area or time restriction is likely to be appropriate. Assuming that it is, however, it must be recognised that there are no simple standard procedures for establishing restrictions. What works for one nation or group of nations can rarely be adopted without modification for a new situation. Nevertheless, one universal truth probably applies - local people must be directly involved in the selection, establishment and management of measures if the chance of success is to be maximised (see Chapter 7). This is also emphasized in the Code of Conduct (see Paragraphs 10.1.2 and 10.1.3).

5.5 Evaluate the need for underpinning research

It is difficult not to be sceptical of some of the bolder claims that are made for the success of area and time restrictions. In particular, with respect to permanent area closures in adult fish habitat, it is certainly not axiomatic that they must enhance fisheries. This is not to suggest, however, that uncertainty about their value as a management tool should be used as a reason not to establish them. On the contrary, the case for trying the permanent reserve approach is rather compelling. However, a manager ought to ensure that we learn from the process, by making efforts to understand the underlying mechanisms that determine success or failure. A program of research that is closely allied with the implementation of the management measure is a pre-requisite for doing this. There are also perhaps more politically compelling reasons for mounting research programs in association with area restrictions. Consider the following scenario.

A marine reserve is set up using arguments that benthic habitats will be conserved and fish stocks will be enhanced. This reserve was established in the face of great resistance by fishers, who perceived the measure as an unnecessary constraint on their trade. In the end the fishing industry accepted the measure, albeit grudgingly. Imagine now that after 5 years, there were no detectable improvements in catches. One could imagine at least four reasons why this might be:

1. the reserve was not big enough;
2. the area had not been closed for long enough;
3. the reserve was in the wrong place;
4. reserves don't work in this system.

Reasons 1 and 2 argue for even stricter constraints, and acting on reason 3 is likely to be politically very difficult. Fishermen of course are likely to argue for reason 4 and for a re-opening of the region. The point is that in the absence of information about the mechanisms that operate in the region, there is no basis for saying which of these explanations is the most likely. Thus, one is unable to decide whether to make the reserve bigger, continue with it as it is, move it, or abandon reserves altogether. Of course, even with a directed research program the information will not be perfect, but adopting the kind of Bayesian approaches described earlier (Section 4.3) seems to be a sensible route towards a sensible decision. Without efforts to monitor the effectiveness of any area or time restriction and understand why they succeed or fail, I fear the goal of protecting fish stocks and the marine system in general will be compromised.

Despite the above comments, it should be recognised that it will often be difficult to demonstrate the benefits offered by an area or time closure for fishery management over reasonable time scales. It has been estimated, for example, that using the standard 5% statistical significance level, it would take over 30 years before one could obtain a 90% probability of recognising a 20% improvement in mean flatfish recruitment after establishing the plaice box described earlier. In many circumstances, therefore the benefit of a closed area is unlikely to be clearly demonstrated in less than a decade (Horwood, 2000).

6. CONCLUDING COMMENTS

There are many compelling reasons why a fisheries manager should seriously consider closed areas and/or time restrictions, either as a complement to other measures or as the primary facet of the management strategy.

Time restrictions have been shown to be effective in many fisheries and are an important tool in the management armoury. In many respects justifications for their use (i.e. the benefits that will flow to fishers) and the process of implementation are likely to be relatively straightforward compared to permanent area closures. The political gulf between a temporary loss of access each year and a loss in perpetuity is enormous.

From a fisheries management perspective, the benefits that flow from permanent area closures are usually less easy to predict than for seasonal closures. Moreover, even if one accepts that implementation of a permanent reserve will provide higher production levels in adjacent fished areas, the potential benefits may often be in danger of being largely dissipated. If, for example, the fishery remains open access, the increased production is likely to attract new entrants into the fishery, thereby driving it back towards bioeconomic equilibrium. From a wider conservation perspective, however, closed areas have an important and clearly defensible role to play and some form of zoning arrangement will often effectively serve conservation values. Given the increasing trend towards the establishment of national networks of marine protected areas, it seems likely that fisheries managers throughout the world will need to ensure that they are familiar with the issues surrounding these approaches.

7. REFERENCES

Anon. 1997. *Towards Sustainable Fisheries. Economic Aspects of the Management of Living Marine Resources.* Organisation for Economic Co-operation and Development, Paris.

Attwood, C.G. & Bennett, B.A. 1994. Variation in dispersal of Galjoen (*Coracinus capensis*) (Teleostei: Coracinidae) from a marine reserve. *Canadian Journal of Fisheries and Aquatic Science*, **51**. 1247-1257.

Costanza, R., d'Arge, R., de Groot, R., Farber, S., Grasso, M., Hannon, B., Limburg, K., Naeem, S., O'Neill, R., Paruelo, J., Raskin, R., Sutton, P., & van den Belt, M. 1997. The value of the world's ecosystem services and natural capital. *Nature (London)*, **387**. 253-260.

FAO. 1996. Precautionary approach to fisheries. Part 1: Guidelines on the precautionary approach to capture fisheries and species introductions. *FAO Fisheries Technical Paper* **350/1**. FAO, Rome.

Fogarty, M.J. & Murawski S.A. 1998. Large-scale disturbance and the structure of marine ecosystems: fishery impacts on Georges Bank. *Ecological Applications*, **8**. S6-S22.

Hall, S.J. 1999. *The Effects of Fishing on Marine Ecosystems and Communities.* Blackwell Science, Oxford. 274pp.

Horwood, J.W. 2000. No-take zones: a management context. *Effects of Fishing on Non-Target Species and Habitats: Biological, conservation and socio-economic issues.* Eds M.J. Kaiser & S.J. De Groot. Blackwell Science, Oxford. 302-312.

Kelleher, G & Kenchington, R. 1992. *Guidelines for Establishing Marine Protected Areas.* A Marine Conservation and Development Report. IUCN, Gland, Switzerland. vii + 79pp.

Lauck, T., Clark, C.W., Mangel, M. & Munro, G.R. 1998. Implementing the precautionary principle in fisheries management through marine reserves. *Ecological Applications*, **8**. S72-S78.

McClanahan, T.R. & Kaunda-Arara, B. 1996. Fishery recovery in a coral reef marine park and its effect on the adjacent fishery. *Conservation Biology*, **10**. 1187-1199.

Russ, G.R. & A.C. Alcala. 1996. Do marine reserves export adult fish biomass? Evidence from Apo Island, Central Philippines. *Marine Ecology Progress Series*, **132**. 1-9.

Murawski, S.A., Brown, R., Lai, H.L., Rago, P.J. & Hendrikson, L. 2000. Large-scale closed areas as a fishery-management tool in temperate marine systems: the Georges Bank experience. *Bulletin of Marine Science*, **66**. 775-798.

Sainsbury, K.J. 1987. Assessment and management of the demersal fishery on the continental shelf of northwestern Australia. In *Tropical snappers and groupers: biology and fisheries management*. Eds J.J. Polovina & S. Ralston. Westview Press, Boulder, CO. 465-503.

Sainsbury, K. J. (1988). The ecological basis of multispecies fisheries management of a demersal fishery in tropical Australia. In *Fish Population Dynamics*. Ed. J. A. Gulland. John Wiley & Sons, Chichester. 349-382.

Sainsbury, K.J., Campbell, R.A., Lindholm, R. & Whitelaw, A.W. 1997. Experimental management of an Australian multispecies fishery: examining the possibility of trawl-induced habitat modification. *Global Trends: fisheries management*. Eds K. Pikitch, D.D. Huppert, & M.P. Sissenwine. American Fisheries Society, Bethesda, Maryland. 107-112.

CHAPTER 4

INPUT AND OUTPUT CONTROLS: THE PRACTICE OF FISHING EFFORT AND CATCH MANAGEMENT IN RESPONSIBLE FISHERIES

by

John POPE

Norfolk, United Kingdom

1.	INTRODUCTION	75
2.	WHAT ARE INPUT AND OUTPUT CONTROLS?	76
	2.1 Input controls or fishing effort management	76
	2.2 Output controls or catch management	76
	2.3 The need for fishing effort and catch management controls to be generally applied	77
3.	WHY WOULD YOU WANT TO USE EFFORT OR CATCH MANAGEMENT?	77
	3.1 How do they link with the objectives of fisheries management?	77
4.	HOW WOULD YOU IMPOSE FISHING EFFORT MANAGEMENT AND CATCH MANAGEMENT?	78
	4.1 Requirement for restrictive licensing	78
	4.2 Reducing fleet capacity	79
	4.3 Forms of catch management	82
5.	WHAT STRUCTURES DO YOU NEED FOR EFFORT AND CATCH MANAGEMENT?	85
	5.1 The centralised nature of fishing effort management and catch management	85
	5.2 Monitoring, enforcement and advisory structures	87
6.	WHAT PROBLEMS EXIST WITH THE APPLICATION OF EFFORT MANAGEMENT AND CATCH MANAGEMENT AND HOW MIGHT THEY BE CIRCUMVENTED?	88
	6.1 The problem of effort management	88
	6.2 Problems with catch management	89
7.	THE PRECAUTIONARY APPROACH AND FISHING EFFORT AND CATCH MANAGEMENT	90
8.	WHERE CAN YOU SEE EXAMPLES OF EFFORT MANAGEMENT AND CATCH MANAGEMENT IN ACTION?	91
9.	REFERENCES	92

1. INTRODUCTION

Fishery resources are limited. Consequently, if fishing pressure is not controlled in some way, it will increase until at best the fishery just breaks even economically and at worst the stock collapses through being unable to reproduce itself. Various forms of management are possible. These are

- technical management (controls on the types of fishing gears allowed and restrictions on times and areas of harvest, see Chapters 2 and 3),

- economic management and social management (see Chapter 5), and

- management of the inputs and outputs to a fishery, the subject of this Chapter. These are the limits on the total intensity of use of the gear fishers put into the water in order to catch fish (fishing effort management or input controls) or the limits on how much fish they can take out of the water (management of catch or output controls). Collectively we sometimes refer to these as "direct conservation measures". This is in order to make it clear to Ministers that these measures, like technical conservation measures, are designed to conserve fish and are not just a way of slicing up the pie! They are essentially concerned with limiting the proportion of fish killed each year by fishing, rather than limiting the sizes, areas and times at which fish are captured.

Conservation of fish stocks is at the heart of the FAO Code of Conduct for Responsible Fisheries (FAO, 1995a) because if the fish do not exist all other objectives fail (Code of Conduct, Paragraphs 6.2, 6.3, 7.1.1 and particularly 7.2.1). Hence limiting the intensity of fishing is a key tool of conservation (Code of Conduct, Paragraphs 7.1.8 and 7.6.1). Consequently this chapter describes how this can be achieved by limiting inputs (fishing effort) and outputs (catch) and explains the requirements and the advantages and problems of these approaches to conservation.

2. WHAT ARE INPUT AND OUTPUT CONTROLS?

2.1 Input controls or fishing effort management

As defined above, input controls are restrictions put on the intensity of use of gear that fishers use to catch fish. Most commonly these refer to restrictions on the number and size of fishing vessels (fishing capacity controls), the amount of time fishing vessels are allowed to fish (vessel usage controls) or the product of capacity and usage (fishing effort controls). Often fishing effort is a useful measure of the ability of a fleet to catch a given proportion of the fish stock each year. When fishing effort increases, all else being equal, we would expect the proportion of fish caught to increase.

For some fisheries, vessels may deploy a variable amount of fishing gear. In these cases the definition of fishing effort would also need to contain a factor relating to gear usage per vessel. In principle, input controls might also refer to limits placed upon other vital supplies of fishing such as the amount of fuel use allowed (energy conservation is desirable, see Paragraphs 8.6.1 and 8.6.2 in the Code of Conduct) but the commonest form of input controls are those put on the various components of fishing effort. In simpler less mechanised fisheries input controls might relate to the number of fishing gears deployed (e.g. the number of static fish traps) or to the number of individual fishers allowed to fish.

2.2 Output controls or catch management

By contrast, output controls are direct limits on the amount of fish coming out of a fishery (fish is used here to include shellfish and other harvested living aquatic animals). Obvious forms of output control are limits placed upon the tonnage of fish or the number of fish that may be caught from a fishery in a period of time (e.g. total allowable catches; in reality, usually total allowable landings). Another form of output control is the bag limits (restrictions of the number of fish that may be landed in a day) used in many recreational fisheries. Limiting bycatch might also be seen as an output control. It is worth immediately noting that to limit fishing intensity it is necessary (unless, as is not usually the case, fish can be released alive) to limit the catch (the amount taken from the sea) rather than the landing (which may well contain only a selection of the catch). The unlanded part of the catch (the discards) may be a substantial proportion of the total catch (Alverson et al, 1994) and may undermine the intent of catch management.

2.3 The need for fishing effort and catch management controls to be generally applied

It is important to notice that neither the management of fishing effort or of catch are likely to be effective unless they apply to all the fishers (or at least the overwhelming majority) engaged in a fishery. Partial controls leave space for the uncontrolled part of a fishery to expand into any gap left by controls placed upon other parts of the fishery. In the past a number of countries only controlled the effort of the larger fishing units on the basis that they created the most fishing pressure. Small vessel sectors of fishing fleets were left uncontrolled since they were thought to take only a small slice of the catch. This resulted in an uncontrolled expansion of the small vessel sector, which modern technology can render very effective at killing fish. Consequently the Code of Conduct encourages managers to take measures for all vessels under their jurisdiction (Paragraphs 6.10 and 7.6.2)

3. WHY WOULD YOU WANT TO USE EFFORT OR CATCH MANAGEMENT?

3.1 How do they link with the objectives of fisheries management?

The quick answer to the title of this section is given in the introduction to this Chapter. It is because fishery resources are limited and if fishing mortality is not controlled, it will increase until the fishery becomes economically non viable or the stocks collapse to extinction (see Chapter 6, Section 2). Thus in most cases the management of fishing effort or of catch are seen as pure conservation measures.

Restricting the amount of fishing by either effort or catch management is one way of protecting fish stocks from becoming over-exploited or of encouraging the recovery of stocks that are depleted as a result of having been over-exploited in the past. So they are one means of achieving the biological conservation of fish stocks. However, as described in Chapter 7, fisheries necessarily involve people and therefore have social and economic as well as biological objectives. Thus, viewing these or other management measures purely as conservation tools is naïve. The social and economic objectives are why people fish and why managers particularly wish to conserve fish. Because of this, the Code of Conduct requires managers to take social and economic factors into account when setting objectives and designing management approaches (see Paragraphs 7.2.2 and 7.6.7). Thus, it is important to consider how the management of fishing effort or of catch may effect the social and economic outcomes of the fishery. By doing this, managers may chose approaches that match the outcomes they desire or at least avoid approaches which would result in undesired outcomes (FAO, 1983; Pope, 1983; McGoodwin, 1990).

Uncontrolled fishing effort tends to increase until, on average, individual fishers make at best only moderate profits and often no profit at all. In unmanaged fisheries this tendency often leads to the fishery becoming biologically over-exploited by being subjected to too much fishing effort and hence to an excessive annual removal rate of fish. This results in fish being caught at a size before they have realised their full growth potential and often before they have an adequate chance to reproduce. This latter tendency is of course far more dangerous. In single-species fisheries the imposition of suitable technical conservation measures may be able to prevent the biological over-exploitation by protecting young fish and/or spawning fish and/or by making the fishery sufficiently inefficient that the zero profit level is reached before the stock is over-exploited. Where the social objective of maximum employment from the fishery is desired this may seem a perfectly sensible approach to management. Just stop fishers doing the things that would lead to over-exploitation (e.g. depleting the spawning population) and then let them fish as much as they wish to. However, this approach may fail if either the costs of fishing decrease (e.g. fishing gear becomes cheaper or more efficient, or fuel becomes cheaper) or the price of fish increases. Consequently the Code of Conduct encourages the limitation of fishing capacity to prevent uncontrolled increases in the amount of fishing (Paragraphs 7.1.8 and 7.6.1).

There may also be problems with a technical management approach if more than one species is targeted in a fishery or if the fishery takes bycatches of vulnerable non-target species (e.g. marine mammals). What may be a suitable technical conservation measure for a species which grows to a small size may be inappropriate for a species which grows to a larger size but which is exploited by the same fishing vessels at the same time. Examples of this are the fisheries for the flatfish species, plaice and sole, caught by the beam trawl fisheries of Northern Europe. Mesh size is smaller than would be optimal for plaice to allow for the catching of the small, lithe but more valuable Dover sole. Thus limiting inputs or outputs from such multi-species fisheries may be a better way of managing them to avoid biological over-exploitation.

If it is desired to maximise the economic benefits of the fishery then other approaches than technical measures are required. If the economic benefit is intended to go wholly to the state then it may suffice to use fiscal measures such as taxation. If set at a level appropriate to the economics of the fishery, such measures can both extract rent from the fishery and also lead fishers to reach the break even/no profit level point with fishing effort that does not over-exploit the stock. However, if it is intended to allow at least some of the profit to devolve to the fishing industry, then it is necessary to stop fishers from increasing their fishing effort at some level short of the zero profit point. It will also be necessary to devise ways to prevent them dissipating their profit by investing in additional uncontrolled inputs, or by engaging in activities such as discarding less valuable fish in order to high grade their landing (Townsend, 1998 discusses new approaches). Hence fishing effort or catch management may be used to secure the biological and either the social or economic objectives of fisheries or some trade-off between biological, economic and social benefits (see Chapter 5). However, how social and economic objectives balance out in practice depends upon the details of how catch or effort restrictions are shared between fishers and this is discussed in Chapter 6.

4. HOW WOULD YOU IMPOSE FISHING EFFORT MANAGEMENT AND CATCH MANAGEMENT?

4.1 Requirement for restrictive licensing

From the previous subsection we have seen that effort management and catch management might serve the biological, economic and social objectives of a fishery. There are a number of ways that effort or catch management may be established but it is the way that they are imposed that will determine which, if any, objectives are satisfied. It is common for countries to require fishing vessels to be licensed. Where they fish on the high seas the UN Fish Stocks Agreement (Article 18) requires the Flag State to control its vessels through licences, authorizations or permits (see also Code of Conduct, Paragraphs 7.6.2, 8.1.1, 8.1.2, and 8.2.1). However, typically such basic registration schemes are not of themselves restrictive, i.e. traditionally a licence might be had by filling in a form and paying a nominal fee. Although such schemes are useful as a basis for statistics and some forms of fisheries control they do not limit the amount of fishing unless coupled to a limited entry scheme.

Clearly, the measures to limit inputs require some form of restrictive licensing which will limit the total number of vessels engaging in a particular fishery together with their fishing power. Often, in order to reduce resistance to restrictive schemes, licensing lists are initially inclusive. They include all vessels, some of which seldom take part in the fishery. Such little used vessels constitute a latent fishing capacity, which might expand its usage to take a larger part in the fishery should it become more profitable. Consequently, it may be wise to extinguish or at least heavily limit these rights if they are not taken up regularly. Failure to do this may force the government to buy out rights when they have become valuable and it is wise to plan for this need when licences are first issued. For the same reason it is also important that the restrictive licence records such characteristics as the size and engine power of the vessel that affect its ability to exploit fish. If these are not fixed then the licence may be transferred to a new, more powerful, vessel or the vessel may be upgraded. Either type of change will allow growth of fishing capacity (Paragraph 7.4.3 in the Code of Conduct suggests studies of these effects).

If restrictive licences are in any sense transferable between owners then they are likely to acquire substantial value to the holder and be transferred at high price. In order for a Government to avoid claims that they are encouraging dangerous practices, typical licence transfer rules must allow for at least limited transfers from an old to a new vessel or from parent to child. The result is that some value is likely to attach to a restrictive licence giving the Government the dilemma that something that they have issued, often for a flat fee, has acquired substantial value that they will probably have to recompense if they wish to rescind the right. In principle this might be overcome by issuing such licences for a fixed term rather than in perpetuity. However, license schemes typically grow from earlier registration schemes and both socially and politically it may be difficult to deny fishers the right to earn a living in a traditional family occupation without offering them compensation. Even where fishing is not a traditional occupation, short-term licences might further discourage fishers from focusing on protecting the long-term productivity of the fish stock. Such issues are discussed further in Chapter 6.

4.2 Reducing fleet capacity

In many cases licensing schemes have been adopted after over-fishing has occurred. In these cases the fleet is already too large. Even where licensing has been brought in early it is quite possible that technological advances in vessel and gear design and improvements in fish-finding and navigation equipment may cause the effective fishing capacity of a fleet to increase through time. Indeed technological improvements in efficiency are often "guesstimated" to increase at about 2% a year. The actual figure may well be higher, particularly if restrictive licensing puts a premium on vessel efficiency. The rules of compound interest mean that even a 2% annual rate of increase in effective fishing capacity will lead to a doubling in fleet capacity in about 36 years; a 4% annual increase leads to a doubling in effective capacity in 18 years. Hence, it is not uncommon for managers to find either that the fleets they are concerned with already have too high a capacity or that they will develop this over time (As an example of this, the over-capacity situation in the European Community is described in the Lassen Report 1996). Thus, if despite licensing, the fleet is too large for the particular fishery then it will be necessary to reduce its capacity (Code of Conduct, Paragraph 7.6.1). This may be arranged in the following ways:

- by removing vessels from the fleet;
- by making all vessels fish for shorter periods;
- by limiting the amount or size of gear that a vessel can carry;
- by reducing the efficiency of fishing effort (e.g. by closing areas where catch rates are high).

Note however that none of these approaches will succeed in reducing the amount of fishing unless entry into the fishery is first ring-fenced (restricted by a limited licensing scheme).

Removing vessels from the fleet

Typically removing vessels from a fleet requires the rescinding of a Government licence. In essence this requires removing a right from an individual for the general good and a just system requires there to be compensation to the owner. Such vessel removals are typically arranged by adopting a Government funded Buy Back or Decommissioning scheme. If the restrictive licences are freely transferable the Government might simply enter the licence market to buy up excess licences. More usually governments announce schemes for fishers to tender for the decommissioning of their vessels or by announcing a price at which they will buy licences.

A general problem with all such voluntary schemes is that the vessels decommissioned are likely to be the least efficient in the fleet. As a result their removal will not cause an equivalent reduction in the ability of the fleet to catch fish. A further problem is that fishing communities are frequently tight-knit and the money paid to an older owner to retire from the fishery may then be recirculated into the fleet capacity, for example by being used to improve the efficiency of a relative's vessel. Indeed as licences become more restrictive there is likely to be a greater incentive to increase vessel efficiency, which may maintain the effective capacity of the fleet and also make successive Government Buy Back schemes more expensive. Thus, there is probably a need to back up such schemes with fiscal measures such as higher licence fees. One rational approach might be to regard Government Buy Backs as an investment loan to the whole industry, which would be funded in whole or in part by subsequent loan repayments from the remaining industry. In general removing vessels from a fleet will tend to increase the profitability of the remaining vessels and thus would serve an economic objective of maximising profit.

Reducing the amount of time vessels are allowed to fish

Reducing fishing time may be arranged by imposing limits on the days vessels may spend fishing. But, once a vessel is over the horizon it may be difficult to check its precise activities! It is true that satellite tracking has the potential to help with defining days fishing but unless special sensors are fitted it can only confirm the vessel was on the fishing ground but not that it was actively fishing (see Chapter 8, Section 3.2.5). Consequently, fishing is usually more practically reduced by limits on days at sea.

Direct restrictions on days fishing are of course possible. A vessel could be given a quota of days during which it might fish (interpreted usually as a number of days that it could not fish and had to be tied up in harbour). Such allowances might be transferable and traded between vessels. In this case they might acquire considerable value if the fishery were profitable. Trading would presumably eventually lead to fleet reductions by all the available days fishing becoming concentrated in an efficient subset of the fleet who could best afford to buy up the rights of other fishers. Thus it is an approach which might tend to generate an economically effective fishery rather than one which emphasized direct employment. As with all effective reductions in fishing effort such schemes may contain the seeds of their own failure by encouraging capital investment in fisheries equipment and/or replacement vessels, which increases fishing capacity. This tendency, called Capital Stuffing, is always present in any input control designed to increase the profitability of fleets for the benefits of their owners. It might be anticipated and counteracted by legislating for days at sea allocations to reduce progressively over time and/or by enforcing reductions in the registered fishing capacity of replacement vessels.

Managers should be aware that unless some transitional compensation is offered, the sudden imposition of days at sea restrictions will be seen by the fishing industry as "decommissioning on the cheap". Such restrictions will usually be resisted since, until the stock responds to lower exploitation, they will reduce or overturn any profit that the industry might hope to make. Days at sea measures (particularly for specified seasons) may be attacked by fishers' groups on the argument that they may encourage people to go to sea to take up their allocation at times when it is dangerous to go out fishing. A general problem with days at sea measures may also be that they often clash badly with fishers' self perception of being free spirits who can go to sea as and when they choose. The success of the implementation of the Faeroese scheme described below was linked to close consultations with the fishing industry.

Other restrictions on time at sea are possible. Where a group of fishers are effectively the sole users of part of a resource they may themselves impose quite heavy restrictions on fishing times. Such restrictions are quite common in European Community fisheries of the Mediterranean where fishers groups (e.g. the Confederes of Spain) impose their own rules. Ports in Catelonia for example fish on a daily pattern and have strictly set hours in which vessels may be at sea. Breach of these cause the vessel to be "fined" additional time the next day.

> The fate of the UK's days at sea restrictions proposed in the UK Sea Fisheries Act 1993 should stand as an object lesson to managers of the problems of imposing a days at sea scheme. Though this Act was passed by the UK Parliament it was fiercely opposed by the collective fishing industry, subjected to Judicial review and to an adverse Parliamentary Select Committee report. Though it remains available for use, the will to impose it was lost. Clearly such schemes are better imposed at the beginning of a fishery when they need not be onerous rather than when real sacrifice is required in the face of severely depleted stocks.

Other time at sea restrictions may be arranged with schemes such as no fishing on weekends. In some cases such rules may accord with local customs and be welcome but in other cases they may discriminate between different groups of fishers. For example a weekend ban might favour small vessel day-boat fishers over those who make more extended voyages. In general reducing the amount of usage of fishing vessels may tend to make the fleet less effective and thus possibly preserve employment, though possibly for shorter working periods. If the time that vessels can be used can be traded between vessels such a restriction might ultimately improve profitability but more slowly than the direct removal of vessels would. As always managers need to think carefully if a particular regulation will have effects beyond the ones they intended or wished (see Code of Conduct, Paragraph 7.6.2).

Restricting vessels use of fishing gear

Some fishing vessels (e.g. otter trawlers) tend to use the size of fishing gear that is appropriate to the vessel's size and horsepower, but even for this gear newer developments (e.g. three bridle twin trawl rig) may increase a vessel's effective fishing power. Restricting the use of such gears may be one way of restricting the increase of the efficiency of fishing effort. For a number of other fishing methods, the amount of gear deployed in a day may have an even less clear relationship to vessel size or engine capacity. This is particularly the case with static gears such as gill nets, pots and creels. The amount of such gears carried by a vessel (or more properly the numbers deployed) may be increased if restrictions are placed upon other aspects of the vessel's efficiency or use. Thus when vessels use fixed gear (e.g. gill nets) days at sea restrictions alone may not be sufficient because fishers may leave out gear that is still fishing while they are in port. Moreover, days at sea restrictions may give an incentive to using more nets or adopting longer gear soak times, in order to maximise the output of catch within the constraint. Such responses by fishers to the legislation may reduce fish quality and may also increase the loss of fixed gear resulting in an increase in ghost fishing. Ghost fishing is caused by lost gear continuing to kill fish (see Chapter 2). In such cases it may be necessary not only to restrict the vessel's capacity and days of usage but also the amount of gear carried. However, this may be a far less easy factor to restrict and manage than vessel days at sea. One approach is to insist that gear is tended by the vessel and lifted when it goes into port. Such restrictions may also be sensible in order to avoid the dangerous overloading of vessels or unsound fishing practices (e.g. leaving gill nets in the water too long; Code of Conduct, Paragraph 6.7).

> In southern Newfoundland where inshore cod fishing is mostly pursued with static gears measures were introduced in the 2000 fishing season to restrict the use of gill nets to the summer season where they were less likely to catch spawning fish or be left fishing or lost due to adverse weather. More generally there seems a movement amongst some inshore fishers in southern Newfoundland to favour a return to more traditional gears such as hook and line and fish traps.
>
> In Bermuda, there is very tight control of the gear used in their fishery for spiny lobster. The government retains ownership of the standard traps, which are the only ones allowed for use in this fishery, and each year leases out a maximum of 300 of these traps to a maximum of 20 licence holders. The traps must be returned to the government at the end of each year for reallocation.

Reducing the efficiency of fishing effort.

Closed seasons are often seen as technical conservation measures and are discussed in detail in Chapter 3. However, they may be seen as a way of restricting effort if their motivation is to reduce fishing time, rather than to affect selection by protecting fish at seasons when certain sizes are particularly vulnerable. Equally the decision to close an area to fishing could be regarded as a *de facto* input control, if it were motivated by a desire to restrict vessel efficiency. Such input restricting closures would be in areas of high catch rates rather than areas where vulnerable ages of fish were found. Obviously, such decisions may well be motivated both by a desire to improve selection (technical conservation) and by reducing the amount of fish removed (direct conservation) and the boundary between these two approaches may become rather fuzzy at this point. Such measures to restrict efficiency will clearly tend to reduce profitability and maintain or increase employment by enabling a greater number of vessels to fish the stock. Such schemes are often adopted by those members of the European Community with Mediterranean seaboards where social objectives tend to dominate.

Effort Management in the Faeroes

The Faeroes, a group of Islands laying between Iceland and the Shetland Isles at the north of the United Kingdom, are part of Denmark but are locally autonomous in most matters including fishing. Fishing is of major importance to the Faeroese economy and an important means of employment. Local tradition favours providing employment for all Faroese but by the early 1990's the need to conserve the local fish stocks particularly those of cod, haddock and saithe was apparent. A total allowable catch system was introduced in 1994 but some fleets misreporting their catches led to the fishing industry rejecting the concept of total allowable catches. Consequently an effort management scheme was introduced in 1996. This imposed limited days at sea, which are transferable within vessel classes, for all but the largest vessels. Larger vessels were restricted from fishing within 12 miles of the coast and a series of area closures was imposed to protect spawning grounds and to reduce the efficiency of fishing. Incentives for fishing beyond the distribution of cod and haddock are also provided by allocating extra days to vessels fishing these areas. There are also bycatch limits of cod and haddock on the larger vessels and on all trawlers. These measures seem to have controlled fishing mortality on the cod and haddock stocks somewhat but have not brought the full reductions intended. They seem generally accepted by the industry as the best compromise possible. Whether the transferable nature of the effort restrictions will lead to concentration of the effort quotas into fewer hands and thus negate the partly social objective of the measure remains to be seen. It also remains to be seen if technical improvements to vessels will require further reductions in fishing effort and if movements of effort between stocks in response to relative abundance generates a virtuous or a vicious cycle of exploitation.

4.3 Forms of catch management

Restrictions on catches may take several forms, the most obvious being the limit on the total catch. This is usually called a total allowable catch (TAC) though catch quota and allowable biological catch are used in some areas. This may sometimes be in terms of numbers of fish (particularly for species that are harvested at a fairly uniform size) but most usually total allowable catches are given in terms of tonnage. Strictly to be effective they should relate to the catch of fish but for administrative convenience they are often limits on landings rather than catches.

The intention of a total allowable catch is to restrict harvest rates to sustainable levels. In reality such restrictions often also provide for the allocation of the resource between user groups. This is particularly important in internationally shared fisheries where some allocation between countries has to be negotiated if a management system is to work. To politicians and to fishers these allocation aspects of total allowable catches may sometimes seem to be more important than the requirements of conservation. Asking how they would share the last fish may put the emphasis back to conservation requirements.

Bag limits are a simpler form of catch limit designed to restrict certain types of fishery by limiting the numbers an individual person or vessel can catch during a short period, typically a day. Such limits do not of course restrict the total catch of the whole fishery. However, they may be effective in restricting sectors such as recreational, first nation or small-scale fisheries that consist of numerous and often dispersed operations that might otherwise be difficult to limit. Such fisheries may have a surprising ability to catch fish. For example recreational fisheries in the United States of America are estimated to take sizeable percentages of a number of species which are also subject to regulation of commercial fishing. One virtue claimed for bag limits is the prevention of recreational fisheries from expanding into semi-commercial operations partially financed by the sale of catch. Another is the prevention of commercial fishing masquerading as recreational fishing; a potential problem when small-scale inshore fisheries are restricted by TAC.

Bycatch restrictions may also be viewed as forms of output control as they restrict either the catch of bycatch species or the proportion that these species form of the total catch, often on a trip by trip basis. The intention of such limits may be to avoid the targeting of depleted species or species protected under some legislation, such as marine mammals. Such rules may need supporting by onboard observers if the restricted bycatch has little or no commercial value. Bycatch limits may also be adopted to restrict the catch of permitted small mesh gears, as far as possible, to those species for which they are intended (e.g. as shrimp). In these cases the limit is placed on the proportion of larger species that may be caught during a trip using the smaller mesh gear. When used in this fashion bycatch limits are adjuncts to technical conservation measures rather than intended as an output control as such. The main output control typically remains the total allowable catch.

The intention of a total allowable catch is to allow sustainable harvest. A TAC should act by restricting the harvest to a safe proportion of the exploitable stock of fish. Therefore to be truly effective, such catches must be related to the size of the exploitable biomass of fish and since this often fluctuates annually, so should the total allowable catch. In practice estimating the size of the exploitable biomass of fish in the sea is a difficult and expensive task (see Chapter 5). It needs be done with reasonable precision because if, for example, the TAC were set at half the exploitable biomass (a not abnormally high proportion) then overestimating this by 100% would in principle allow the whole stock to be caught with disastrous implications for conservation.

Sometimes TACs are set on the basis of average abundance of the stock, and if a stock is relatively lightly exploited this may suffice. However, the effect of average TACs (sometimes called precautionary[1] TACs) will be to harvest a larger proportion of the stock when it is smaller and a smaller proportion when it is larger. This is of course, the wrong way round since the TAC would unduly restrict fishers' activity when the stock was large, and may fail to protect the stock adequately when it is small. Total allowable catches are thus a far more effective form of management if they can be modified periodically to accord to the size of the stock. However, if the exploitation rate is low and/or the stock is not very variable, then this may not need to be done annually. However, more commonly TACs will need to be adjusted annually.

Having set a total allowable catch there is a need to control the fishery so the TAC is not exceeded. A number of approaches are possible, which include:

---

[1] Precautionary in the sense that they are adopted as a default and certainly not to be confused with the ideas of precautionary management.

- free fishing until the total allowable catch is taken and then shutting the fishery down;
- allocating catch by period and then shutting down the fishery for the remainder of each period when the allocation is caught;
- allocating proportions of the TAC to various sectors and leaving them to manage their own share themselves;
- allocating proportions of the TAC to individuals or individual vessels.

Allowing free fishing until the total allowable catch is taken is the simplest way to administer a total allowable catch since it is only necessary to keep a running total of the overall catch and then to stop the fishery when it is exceeded. However, such a fishery will encourage an intense race to catch fish before the total allowable catch is taken. It will thus encourage excessive capacity and a poor economic performance in the fishery. The classic example of this form of management was in the earlier management of the Pacific halibut fishery where the annual season progressively reduced to a very short period before the season closed. Such intensive races may exclude some fishers who are otherwise engaged during the short open season. On the plus side such management may suffer less from discarding practices because fishers probably find it better to land everything. However, an intense race to fish is likely to make fish quality a secondary consideration to landing fish quickly. In short, such a system does not have much beyond simplicity to recommend it.

Options for TAC management

Allocating catch by period and allowing free fishing during each period has some of the advantages of simplicity of the previous approach, in particular the need only to monitor overall catch. It would still encourage a series of races to fish but it would have some advantages in that fish would be landed throughout the year. Indeed the period chosen for opening might be aligned with times when fish quality or fish prices were best. It is however still likely to encourage excessive fishing capacity to develop and thus to be economically inefficient.

Allocating proportions of the TAC to various sectors that manage the catch uptake themselves may encourage a more orderly uptake of the fish. To what extent this occurs will depend upon the sector's ability to govern itself. In some cases producer organisations or co-operative organisations can manage fish uptake between their members very successfully. In other cases where the allocation is to an industry sector such an allocation may just engender a within-sector race to fish. A problem for management is that catch statistics will have to be collected separately for all sectors and moreover sectoral allocations would give each sector some incentive to under-report catch if they feel that they could get away with it or if they suspected other sectors were also cheating. A further problem may be to decide how to handle a case where one sector is found to have exceeded its allocation. Should management take it from the next years allocation of that sector? Alternatively, should they penalise the transgressors but balance the excess catch by reducing the remaining in-year allocations of other sectors? If the TAC is shared between several countries the latter course may be the only one to take but will be deeply unpopular with the other sectors. In better managed sectors there may be a tendency to try to improve quality and price but this may also encourage high grading. This is the practice of discarding lower valued catch in favour of its higher valued portions. Since there are often differential prices by size and since at times fish may be caught in poor quality (for example from having been too long in a gill net) there is considerable temptation for this practice to arise if fishers have a known allocation of catch.

Allocating proportions of the TAC to individuals or individual vessels has the virtues ascribed to the previous approach. It is more likely to lead to an orderly uptake of the TAC and in a fashion likely to be economically efficient. However, such a scheme is particularly vulnerable to administrative drawbacks. Statistics obviously have to be kept at an individual fisher or vessel level and the temptation and the opportunity to misreport increases as the allocations are more finely divided. Such problems of reporting often lead governments to develop rather draconian measures to ensure compliance by individuals. Problems of high grading are also likely to be exacerbated when quotas are organised at an individual level.

As with fishing effort management, catch management can clearly affect the outcome of the various objectives of fisheries management. The proportion of the stock it is intended to remove annually will affect the biological objectives of yield maximisation, yield stability and conservation. The proportion removed may also affect the overall profitability because the potential for achieving maximum profit will occur at exploitation rates less than those that achieve maximum yield. These are also lower than the exploitation rate where the break-even occurs between earnings and costs; the point where more fishers would be employed in an unsubsidised fishery. How the TAC is allocated to individual fishers will also impact the achievement of objectives. As an example of this, a quota might be divided into many small slices designed to provide individual fishers with a viable living. Clearly such slices should not be transferred. Such a non-transferable quota would tend to provide maximum participation in the fishery. By contrast a transferable quota would be traded and tend to aggregate into fewer but more profitable enterprises. This is the management approach that many fisheries economists advocate because they tend to believe in profit maximisation as the best objective. Notable examples of the use of Individual Transferable Quota (ITQ) management schemes are found in New Zealand and Iceland. Increasingly, analogous schemes are to be found in other countries though not always in such a clear form. For example the United Kingdom has quotas tied to producer organisations rather than vessels. By contrast Ireland and Namibia use non-transferable quotas to encourage wider employment opportunities. These approaches are discussed further in Chapter 5.

5. WHAT STRUCTURES DO YOU NEED FOR EFFORT AND CATCH MANAGEMENT?

5.1 The centralised nature of fishing effort management and catch management

By their nature, effort and catch management have to span the whole fishery (Code of Conduct, Paragraph 7.3.1). Small sized resources might be managed on a local basis but the larger, more extensive, resources have to have their management agreed centrally. Since the more important resources tend to be the extensive ones, approaches to both fishing effort management and catch management tend to be centralised command and control approaches. There is usually relatively little ability, except on small local stocks (e.g. the Thames herring), to devolve the decision-making to the districts where fishers operate. A corollary to this is that they will not work well where the central body has little control over the regions where the fisheries actually operate (see Paragraph 7.7.1 of the Code of Conduct). Moreover, their operation may often seem remote from the interested parties, and to fishers the controls may appear driven by the whims of remote bureaucrats who they perceive, sometimes with justification, as being ignorant both of the fish and the fishing industry. Any attempts to put a more human face on the operation and to increase its transparency (Jentoft and McCay, 1995) is thus to be welcomed (see Chapter 7 of the Guidebook and Paragraphs 6.13 and 7.1.9 of the Code of Conduct). A further problem is that such management tools may become the shuttlecocks of political debate. Since this may focus attention on the short rather than the long-term goals of the fishery it may work against the goals of conservation and optimal long-term use. Hence, there may be some virtue in moving the decisions and operation of effort management or catch management out of direct political control and into the care of a benign and transparent public body which is given clear objectives by politicians but is then left to do its job.

> **The Canadian FRCC**
>
> The Fisheries Resource Conservation Council (FRCC) was created in 1993 to form a partnership between scientific and academic expertise, and all sectors of the fishing industry. Together, Council members make public recommendations to the Minister of Fisheries and Oceans of Canada on such issues as total allowable catches (TACs) and other conservation measures for the Canadian Atlantic groundfish fishery. The Council is responsible for advising the Minister on Canada's position with respect to straddling and transboundary stocks under the jurisdiction of international bodies such as the Northwest Atlantic Fisheries Organization (NAFO). The Council also provides advice in the areas of scientific research and assessment priorities.
>
> The Council consists of 15 members, appointed by the Minister of Fisheries and Oceans, with an appropriate balance between 'science' and 'industry'. Members are chosen on merit and standing in the community, and not as representatives of organizations, areas or interests: 'science' members are drawn from government departments, universities or international posts, and are of an appropriate mix of disciplines, including fisheries management and economics; and 'industry' members are knowledgeable of fishing and the fishing industry, and understand the operational and economic impacts of conservation decisions. Members appointed from the Department of Fisheries and Oceans serve 'ex officio'. The four Atlantic Provinces, Quebec, and Nunavut may each nominate a delegate to the Council.

The box gives the example of the Canadian FRCC, which is described as an independent body, which provides advice on fisheries management. Such bodies are also well suited to educating interested parties on conservation issues as some members share some common backgrounds and experience as fishers (see Code of Conduct, Paragraphs 6.16 and 7.1.10).

Where fisheries straddle political boundaries or occur in international waters there will be the need for some intergovernmental body where fisheries management can be discussed and agreed. The Code of Conduct is particularly detailed in this area (Code of Conduct, Paragraphs 6.12, 6.15, 7.1.4, 7.3.4). There will almost certainly have to be some corresponding scientific body to provide agreed scientific advice on the management needs (see Code of Conduct, Paragraphs 7.3.4, 7.4.1). Such bodies exist for most international fisheries. For example NAFO regulates fisheries that occur wholly or partly in its convention area outside the national EEZs of the seaboards of West Greenland, Atlantic Canada, France (St Pierre, Miqelon) and the USA (New England) but multinational fisheries may also be arranged by formal or informal bilateral or multilateral bodies (e.g. Argentina Uruguay joint management of Rio de la Plata fisheries, Norwegian and European Community joint management of the North Sea).

Where fisheries are wholly national affairs they are usually the responsibility of Government Departments. While the sole ownership of the fishery resource may ease some management problems, the problems of national management are often a microcosm of the problems of international management. Regional governmental bodies often have disparate objectives and different fleets, and may well hold different viewpoints on what are the appropriate fisheries controls (Code of Conduct, Paragraphs 6.12 and 7.6.5). This can lead to disagreements both between different regions and with the central government. In these circumstances it is not uncommon for agreement to be reached for political rather than conservation reasons. This is particularly the case since the time horizon of fish stock recovery is often longer than the time for which politicians are elected. This emphasises the desirability of taking management decisions out of the political arena and allowing some involvement of interested parties in the process. What perhaps is required is a fisheries equivalent of the former Bundes Bank of the Federal Republic of Germany which might protect fish stocks and coastal resources in the same way as the latter protected the value of the German currency (see the Code of Conduct, Paragraph 10.1.3).

5.2 Monitoring, enforcement and advisory structures

However management is arranged, there will clearly be a need to collect data from the fishery in order to monitor the compliance of fishers with effort or catch restrictions and to enforce the management of the fishery in some way (see Chapter 8 for detailed discussion). This can seldom be left to the fishing industry and consequently the Government frequently has to act as the referee and to arrange for the fishery to be policed. As a result, there usually has to be some national data collection and inspection service to monitor and to control the fisheries (Code of Conduct, Sub-Article 7.4 for data and Paragraphs 7.1.7 and 7.7.3 for control and surveillance).

Monitoring and enforcement services may be combined since they need similar access to the fishery and monitoring data provide the essential information on non-compliance with fishing effort and with catch restrictions (Chapter 8, Section 1). To be effective they need to be staffed with personnel who understand the workings of and are sympathetic to the problems of the fishing industry but can deal with them fairly yet firmly when the needs arises. In many cases dockside monitoring and enforcement will not fully meet the needs of either fishing effort restrictions (gears used, areas fished and species caught may all need to be checked at sea) or catch restrictions (discarding at sea may need to be monitored or prevented). Hence monitoring and enforcement staff will usually need to be capable of working at sea and for senior staff it may be appropriate that they hold some appropriate form of mariners certification. Most of all they must be incorruptible, which argues that their income must be adequate and they must be carefully selected.

Most fisheries management approaches need some form of scientific advice in order to make informed decisions (see Chapter 5 of this volume, and the Code of Conduct, Paragraphs 7.4.1-5 and all of Article 12). This is especially the case for fishing effort management and catch management. Both are intended to limit the proportion of the fish stocks being removed each year to a sustainable level. Ideally this requires ongoing information on what proportion of the stock is being removed each year. It also requires a reasonable knowledge of what levels of removals are likely to prove sustainable. These facts allow adjustment to be made either to the fishing effort being applied to the stock or to the catch being removed from it, so as to lead to the appropriate proportion being removed in the future. These requirements are in fact quite difficult to provide since they require the enumeration of a resource that cannot be directly seen or counted and an understanding of how it will react to exploitation.

In the case of catch management, estimates of the stock size typically need to be known rather precisely. This is because if it is wildly inaccurate the catch restriction might even be set higher than the size of the stock. At best this would cause no restriction to the fishery and at worst might endanger the stock. Input controls also will need scientific advice of a reasonable precision to set them at an appropriate level initially and to respond to changes in the efficiency of fishing effort through time. However, typically the demands of precision in population estimates for output controls are not so high as for catch management since the removal rate generated by current fishing effort levels may be judged over a series of years rather on a year by year basis. Hence some averaging out of wild results occurs. Where these dubious results do not apply it may be possible to find a measure of relative removal rate and also find what levels of this measure are sustainable. However, the commonest measure of relative removal rate is fishing effort, which is subject to trends in efficiency, which may obscure changes in the stock. Thus a certain level of fishing effort might seem sustainable, judged on the evidence of past catch-per-unit-effort levels, but increasing efficiency might be masking a decline in the stock. Therefore, fisheries independent measures of stock abundance such as scientific fishing surveys may provide a safer measure of stock status than catch rates based upon commercial fishing effort.

Advice on stock status requires some suitable source of scientific advice (Chapter 5). This might possibly be acquired on an ad hoc basis from university departments but if advice is more than minimal it will probably require either a dedicated scientific agency or the ability to commission appropriate scientific advice. The provision of advice also typically requires some scientific as opposed to management data collection since scientific assessment of a stock typically requires a more detailed understanding than is required for routine monitoring, control and surveillance.

In addition to the biological advice which underpins the setting of limits it will also be wise to have advice on the likely economic and social effects of the fishery. Since these may well give conflicting signals, such advice is best integrated so that the trade-offs between each effect can be clearly seen (See Code of Conduct 7.4.5).

6. WHAT PROBLEMS EXIST WITH THE APPLICATION OF EFFORT MANAGEMENT AND CATCH MANAGEMENT AND HOW MIGHT THEY BE CIRCUMVENTED?

6.1 The problem of effort management

A major problem of effort management is defining a reasonable unit measure of fishing effort. For example, for towed fishing gears (i.e. otter trawls, dredges, beam trawls) some combination of vessel tonnage and engine power often seems appropriate as a measure of the ability to catch a proportion of the stock. As an example of this, the European Community effort management system (called the Multi-annual Guidance Programme) uses both tonnage and also engine power to define alternative measures of fishing power, both of which must decrease through time to agreed schedules (for a simple general description of the Common Fisheries Policy, which uses both effort management and catch management, see European Commission, 1994). For static gear fishing methods other effort measures might be more appropriate as a consistent indicator of fishing power but for some methods no such consistent measure exists. This is because a major requirement of any input control is that a given amount of the input regulated should correspond to a constant ability to exploit fish. This tends to be the case for demersal towed gear fisheries. However, for fisheries on schooling pelagic species the proportion of the fish stock caught by a given unit of fishing effort may vary depending on the size of the stock. If the number of schools of fish diminish as the stock size diminishes then it will become progressively easier to take larger proportions of the stock as its abundance declines. Such stocks are clearly not suited to management by input controls.

Where effort management approaches are appropriate then the main problems centre around the increases in technical efficiency that tend to occur as time goes on. The problem of capital stuffing has already been mentioned. It is rather obvious that if fishers cannot expand a profitable operation by increasing the size or number of their vessels they may try to do so by spending more on capital improvements designed to increase its efficiency. Whether there is an upper limit to possible efficiency improvements is a moot point. However, if there is a theoretical upper limit it is clear that it has not yet been reached since efficiency still tends to increase. Certainly fishing gear technology has advanced inexorably over the last century and with it the ability of a given size of vessel to kill a greater proportion of the fish available. Consequently restricting other outlets for fishers' investment by effort management is likely to make technical improvements in vessel efficiency very tempting. Either removing super normal profits by fiscal measures or progressively reducing the fleet capacity seem the most appropriate responses to this. The alternative of specifying fishing vessel characteristics as tightly as those of a class of racing yachts seems inappropriate and stifling of innovation. Moreover given the ingenuity of mankind it is unlikely to work since fishers will find ways to improve any unregulated dimensions of the inputs. Thus it is most important to anticipate that such efficiency improvements will occur and to anticipate how they will be handled when they do.

A particular problem is that licences may frequently specify engine power. Engines may readily have their power output derated by minor technical modifications. Such engines can thus be certified to comply with a licence requirement but after certification their performance may be subsequently subtly upgraded! Periodic certification and inspection may be needed to close this loophole.

Other problems concern non-licensed fishing. This may be a particular problem for fisheries in international waters where fleets flagged in countries that are not part of the relevant international organisation may feel inclined to disregard its decisions. They most certainly should not do so and the Code of Conduct is very clear on this point (see Paragraphs 6.1, 7.1.5, and 8.2.6). The FAO Compliance Agreement and the IPOA on Illegal, Unregulated and Unreported Fishing (Chapter 1 Table 2) both address this important matter.

In multi-species fisheries there is an obvious risk that fishing effort may switch between the various available fish stocks depending on which is currently the most profitable. To some extent this might be beneficial since fishing effort might tend to move off of stocks which are at low abundance and move onto ones which are at high abundance. However, there remains the risk that the more valuable species will tend to attract more of the fishing pressure than less valuable species. Moreover, scarcity might cause particularly valuable species to increase in unit price sufficiently to continue to attract effort even when they are over-exploited and at low abundance. Such problems might best be addressed by also imposing technical conservation measures designed to protect the more valuable and more vulnerable species of an assemblage.

A further problem may occur where several different fisheries coexist. In such cases it may be difficult to ensure that a vessel licensed to participate in one fishery is not in fact participating in another.

6.2 Problems with catch management

As discussed earlier with simple output control approaches, (when the fishery is closed just as soon as the quota is taken) a race to fish is almost bound to develop and may be associated with over capacity, poor economic performance, poor fish product price and quality, and only seasonal employment (see the Code of Conduct, Paragraph 6.7). Moreover, a rush to catch fish may cause the safety of crews to be compromised. With more orderly management of quotas with allocations made to groups or individuals these problems are less likely to occur. The major problem with such a form of output control is the frequent non-compliance or circumvention of the regulation. In many fisheries there is bound to be an economic temptation for fishers to land more fish than their allocations allow. Such illegal landings are often called "black landings" (Alverson et al, 1994). If they are extensive they may undermine faith in the management process (see box on Faeroes fisheries management). Moreover, the distortion of catch statistics that they may create may also make it difficult for fisheries scientists to provide accurate estimates of future catches. The reduced ability of scientists to predict catches may lead to a further lack of faith in the management system. The obvious remedy is tighter enforcement coupled with an education program to make fishers realise that they are cheating fellow fishers rather than the Government. This is likely to be facilitated by a more open system of governance that is not too close to that of Central Government and has industry representatives as members.

A more subtle form of non-compliance is reporting one quota species as another species or reporting a species as coming from another management area. Such forms of non-compliance are sometimes referred to as "grey landings". The consequences of grey landings in terms of reduced faith in the system can be as pernicious as those of "black landings". Indeed they may be worse since they foul the statistics of at least two stocks. Again proper regulation and education are the appropriate responses to prevent this happening.

Quota regulations are often specified as levels of legal landings rather than catch. So fishers may chose to discard some of the less valuable part of their catch in order to get the best earnings from the quota they have available. Such discarding is often legal and indeed fishers may be required to discard over-quota or undersized fish. Such discarding is wasteful and should be minimised (see the Code of Conduct, Paragraphs 6.7, 7.6.9 and 8.4.5). Moreover, such discarding practices may also distort the intention of the regulation and allow exploitation to be higher than the regulation intended. Again this is a practice that may undermine faith in the management system. The prevention of discarding presents a dilemma. To permit discarding allows it to be measured by scientific observers (though this is expensive), to ban it may discourage it but also may simply render it unobservable. Some countries (e.g. Norway) require all fish to be landed and some compensate fishers for the cost of landing illegal or unmarketable fish. Compensation is set at a level that encourages landing but provides no real profit to accrue to fishers from catching the fish. Other approaches involve the setting up of permanent or temporary closed areas to direct fishers away from the areas where fish that are likely to be discarded are abundant (Code of Conduct, Paragraph 7.6.9). Temporary closed areas obviously require fast footwork by managers to be effective and are thus more easily achieved when the industry is involved in the details of the management and can provide rapid intelligence. (Temporary closed areas are used, for example, in Norway and Iceland, and the pelagic fishing industry in South Africa imposes closed areas on its members as a means of reducing bycatch).

Discarding can be a particular problem in multi-species fisheries. When a quota for one species is exhausted, fishers may continue to catch it and to discard it while fishing for other species in the area. This creates a particular problem if one species is recovering from overexploitation and quotas have been set on the basis of low removal rates, while the fishery for another species allows far higher removal rates. The problem is of course exacerbated if both species are caught with the same fishing gear at the same times and in the same places. Allowing the limited landings of bycatches of the more vulnerable species may be one management approach but this may cause a fishing pattern to develop which utilises any bycatch provision to the maximum extent possible. Suitable technical measures may be another approach (Code of Conduct, Paragraph 7.6.9) but the problem may be that the vulnerable species is larger and more selected by fishing gears than the less vulnerable species.

A draconian but effective approach is to close the entire fishery once one quota is exhausted or once some specified bycatch level has been exceeded. This is practised in the USA's Bering Sea and Gulf of Alaska fisheries to protect Pacific halibut and marine mammals. It requires a dedicated observer programme for such an approach to be effective. Where observer programmes do not operate, there is a risk that some fleets might make anticipatory discards of any species with quotas or bycatch levels, which if exceeded would lead to a closure of the total fishery.

7. THE PRECAUTIONARY APPROACH AND FISHING EFFORT AND CATCH MANAGEMENT

The precautionary approach to fisheries management (FAO,1995b, FAO, 1996) can be applied at all stages of development of fisheries (Code of Conduct, Paragraphs 7.5.1 and 7.5.2). It is not particularly associated with any one management approach. However, as fish stocks become progressively more exploited it is likely they may require the reduction in exploitation rates that effort management or catch management are intended to provide. An intelligent anticipation of this need may certainly help. This could be achieved by adopting restrictive licensing schemes and detailed catch statistics before the need for them becomes acute. Such intelligent anticipation of future problems by the development of a suitable fisheries management plan (see Chapter 9 of this volume and Code of Conduct, Paragraph 7.3.3) is an integral part of the precautionary approach. More detailed interpretation of the precautionary approach becomes possible once a full range of effort management or catch management measures are in place. In particular the imposition of target and limit reference points for a fishery may be made more possible if the population dynamics of a stock are known in the detail often required by catch management (Code of Conduct, Paragraph 12.13).

Detailed scientific thinking about recovery plans compatible with the precautionary approach has been developed (ICES, 1997). These typically involve progressive reductions in fishing mortality rate as stock sizes diminish. Such detailed management probably requires effort management and/or catch management for their achievement. However, the alternative or addition of safe technical conservation measures, for example towed gear mesh sizes that give fish a chance to spawn before they are caught, or alternatively extensive no take zones might help prevent the stock becoming seriously depleted in the first place.

8. WHERE CAN YOU SEE EXAMPLES OF EFFORT MANAGEMENT AND CATCH MANAGEMENT IN ACTION?

Catch management, notably total allowable catch (TAC) management systems, is particularly common in fisheries where the catch is based upon fewer species. In the case of demersal (bottom living) fisheries these are more often found in higher latitudes. Most of the major demersal fisheries of the European fisheries of the North Atlantic have total allowable catches as the primary management instrument. The same is true of the demersal fisheries of Canada and much of the USA. Total allowable catch management is also practised in Argentina, Australia, Namibia, New Zealand, and South Africa. One may also find them used on pelagic fisheries which tend to be less mixed than demersal fisheries. Notably, they are used on large pelagic fisheries such as the southern bluefin tuna.

Total allowable catches have an obvious appeal when fishing opportunities have to be shared between countries, or communities or fleets since they can be allocated a constant share of the overall TAC. Such percentage shares of catch form the basis of many fisheries agreements between countries. In general it is easier for countries to agree to share catch in some proportion than to agree how to share out fishing effort. This is because fishing effort is measured in various ways for various fleets so that establishing a common currency for an agreement is technically difficult.

Moreover, it is well known that fishing effort may change in efficiency through time and changes in efficiency by one partner might well undermine any agreement. Thus percentage shares of catch (so called relative stability) form the basis of national shares of the European Community's Atlantic fisheries. It is also the basis of the share between Norway and the European Community in the North Sea and Norway and Russia in the Barents Sea and between Australia, Japan and New Zealand for the southern bluefin tuna. For these reasons total allowable catch management remains established in these areas even where its track record for achieving sustainable fisheries is anything but impressive, e.g. in the demersal fisheries of the European Community.

While the use of total allowable catch management is widespread in higher latitude demersal and pelagic fisheries, it appears to be more difficult to operate as the number of species increase in a given fishery. There are several reasons for this. Firstly, problems of bycatch are bound to increase as more species occur in the catch and it is more likely that several of the quotas managing such a fishery will be incompatible and lead to discarding or falsified landing declarations. Secondly the requirement for scientific assessments of stock sizes becomes more difficult and less cost effective when faced with many small stocks rather than one big one. This is because the amount of sampling needed to effectively sample a small stock is similar to that required for sampling a large one. The cost of scientific advice may thus tend to be lower per unit catch for a large fish stock than a small one. The same is often true of the per-unit costs of management monitoring and surveillance. Therefore, by extension, the species rich demersal fisheries of tropical regions may be virtually impossible to manage by single-species quotas. The alternative of multi-species quotas of course always carries the risk that the fishery may focus on the more valuable species and perhaps discard the less valuable species in order to maximise short-term earnings.

Effort management systems have a less systematic distribution. Clearly they may be inappropriate for managing fisheries on schooling pelagic fish whose catchability may increase as stock sizes decrease. They seem appropriate for some single-species fisheries particularly where precise scientific assessment of stock size is difficult but where a reasonable presumption of constant catchability exists. Examples of this are seen in some local shellfish fisheries.

Mozambique deep-water shrimp fishery: this fishery illustrates the potential advantages of effort management but also the problems of success. The fishery on Sofala bank catches two species of shrimp. One species has continuous recruitment while the other recruits to the fishery during November and December. An industrial trawler fishery started in the late 1970s and rapidly became the most valuable fishery in Mozambique. It was mainly fished by two joint venture fleets from which Mozambique earned licensing fees. However, the stocks became over-exploited by the early 1990s. A limited entry scheme and a total allowable catch were introduced but the latter was set at too high a total allowable catch to form a binding constraint on the fishery. The Mozambique Fisheries Research Institute (IIP) proposed a closed season for January and February both as a technical measure and also as an effort reduction. This resulted in a rapid recovery of the fishery to profitable levels. The closed season was subsequently extended to also include December. However, the profitable nature of the fishery led to the issue of additional licences and the fishery is again depressed. For details see www.mozpesca.org

---

Effort management seems appropriate for multi-species trawl fisheries where the ability of the fleets to move their attentions to the most abundant species may help reduce pressure on depleted stocks. However, this is not an automatic benefit. It would certainly depend upon the fleets not being so large that they exhausted one stock before the rested stock recovered. It would also depend upon how separate the distributions of the different fish were and how their relative prices adapted to scarcity or abundance.

Effort management is also used as a backstop for fisheries predominantly managed by TAC. An example is the Atlantic fisheries of the European Community (see European Commission, 1994 for a broad outline of the Common Fisheries Policy and the Lassen Report, 1996 for the specifics of over capacity) where the capacity of the fleet is out of balance with the resources and needs to be reduced. In the case of the European Community an effort management programme called the Multi-annual Guidance Programme has been in place for some time. However, it has suffered in that the annual reductions that European Community countries could agree are insufficient to stem the typical increases in the efficiency of fishing effort that may be expected.

Experience suggests that single tools seldom suffice to achieve fisheries management. Moreover, where multiple objectives exist, multiple management tools will certainly be required (Pope, 1983 and Chapter 5 of this volume). However, it is also true that the historical problems with both fishing effort management and catch restrictions were that concerns for short-term goals often clouded the achievement of long term goals. Consequently fishing effort and catch management have not historically been applied with sufficient vigour or with sufficient regard to the precautionary approach to achieve long-term sustainability. Achieving long-term goals by input and output controls or other methods will require an appropriate focus and considerable political will.

9. REFERENCES

Alverson, D. L., Freeberg, M. H., Murawski, S. A., and Pope, J.G. 1994. A global assessment of fisheries bycatch and discards. *FAO Fisheries Technical Paper*, **339**. FAO, Rome. 233pp.

European Commission. 1994. *The New Common Fisheries Policy*. Office for Official Publications of the European Communities, Luxembourg. 46pp.

FAO. 1993. Report of the Expert Consultation on the regulation of fishing effort (fishing mortality). A preparatory meeting for the FAO World Conference on fisheries management and development. Rome, 17-26 January 1983. *FAO Fisheries Report* **R289**. FAO, Rome. 34pp.

FAO. 1995a. *Code of Conduct for Responsible Fisheries*. FAO, Rome. 41pp.

FAO. 1995b. Precautionary Approach to Fisheries. Part 1: Guidelines on the Precautionary Approach to Capture Fisheries and Species Introductions. *FAO Fisheries Technical Paper* **350**/1. FAO, Rome. 52pp.

FAO. 1996. *FAO Technical Guidelines for Responsible Fisheries No. 2. Precautionary Approach to Capture Fisheries*. FAO, Rome. 54pp.

ICES. 1997. Report of the Study Group on the Precautionary Approach to Fisheries Management, ICES Headquarters, 5-11 February 1997.ICES CM 1997/Assess:7. 41pp.

Jentoft, S and McCay, B. 1995 User Participation in Fisheries Management: Lessons drawn from international experiences. *Marine Policy*, **19**: 227-246.

Lassen Report. 1996. Report of a Group of Independent Experts to advise the European Commission on the Fourth Generation of Multi-annual Guidance Programmes, Commission of the European Communities. DG XIV Brussels.

McGoodwin, J.R. 1990. *Crisis in the World's Fisheries: People , Problems and Policies*. Stanford University Press, Stanford.

Pope, J. G., 1983. Fisheries Resource Management Theory and Practice. In: Taylor, J. L. and Baird, G. G. (Eds.) *New Zealand finfish Fisheries: the Resources and their Management*. Trade Publications, Auckland. p. 56-62

Townsend, R.E. 1998. Beyond ITQs: Property Rights as a Management Tool. *Fisheries Research*, **37**. 203-210.

CHAPTER 5

THE USE OF SCIENTIFIC INFORMATION IN THE DESIGN OF MANAGEMENT STRATEGIES

by

Kevern L. COCHRANE

Fishery Resources Division, FAO

1. INTRODUCTION 96

2. WHAT DATA AND INFORMATION DO I NEED? 97
   2.1 What information is needed to help make a decision? 97
   2.2 Where do I get the information and how can I use it? 99

3. HOW MUCH FISH SHOULD BE CAUGHT: HARVESTING STRATEGIES AND REFERENCE POINTS? 99
   3.1 Basic harvesting strategies 101
   3.2 The classic reference point : maximum sustainable yield 102
   3.3 Reference points based on fishing mortality rate 103
   3.4 Reference points based on size-limits 105
   3.5 Multi-species and ecosystem-based reference points 105
   3.6 Economic and social reference points 106

4. WHAT TOOLS CAN BE USED TO GENERATE INFORMATION TO ADVISE MANAGEMENT? ...106
   4.1 Single-species methods 107
   4.2 Multi-species methods 109
   4.3 Considering the benefits to society 111

5. HOW IS THE INFORMATION USED TO DEVELOP A MANAGEMENT STRATEGY? 111
   5.1 What sort of biological information is needed? 111
   5.2 What sort of ecological information is needed? 115
   5.3 What sort of social and economic information is needed? 117

6. THE ROLE OF THE SCIENTIST: PROVIDING OBJECTIVE INFORMATION 119

7. HOW SHOULD DECISION-MAKERS AND THE PROVIDERS OF INFORMATION WORK TOGETHER? 120

8. PRESENTING INFORMATION TO DECISION-MAKERS 120

9. WHAT ABOUT UNCERTAINTY? 123

10. UNCERTAINTY AND THE PRECAUTIONARY APPROACH 125

11. CONCLUSIONS 125

12. RECOMMENDED READING 126

APPENDIX: RISK ASSESSMENT 128

1. INTRODUCTION

A typical marine ecosystem is a dynamic and complicated network of natural populations, sometimes spread over tens of thousands of square kilometres, continually changing and moving, influenced by the variable and usually unpredictable meteorological and marine environments. The fisheries exploiting those natural populations are a part of the ecosystem and are also complex and dynamic, using gear-types, fishing strategies and expert knowledge that differ from fisher to fisher or vessel to vessel and are also likely to change with time. To make it even more difficult, the fish and invertebrate populations are usually widely dispersed, hidden from our view and hence very difficult to monitor. Their growth and mortality rates can and usually do change considerably with age and over time and recruitment of young fish to each stock is highly variable. With all of these complexities and uncertainties put together, the fisheries manager is operating in a complex and confusing environment. However, livelihoods and incomes depend on wise decisions made by the managers, and wise decisions are only possible if the managers have adequate knowledge of the ecosystem and fishery to allow them to understand the causes of the current situation in the fishery and to forecast how the resource and fishery will change in response to management actions. The purpose of this chapter is to examine the issues which need to be considered by the manager in implementing effective management strategies, the information which the manager should attempt to have available to guide those decisions, and how this information should be used in making them.

The fisheries manager is likely to be, and should be, involved in setting the fisheries policy and goals (Steps 1. and 2., Table 1). Policy and goals are a part of the strategic planning of the fishery and are usually put in place and modified infrequently, typically being reviewed only every five years or longer (see also Chapter 9). They set the framework for the fishery during this period and should be established with careful consideration of the best available knowledge of the resources and fishery. On a day to day basis, however, the manager is likely to be more involved with the shorter term, tactical decisions of fisheries management, translating the goals into operational objectives and ensuring that the management strategy being used is the best means of achieving those objectives. These are primary tasks of fisheries management and this chapter focuses on how the manager should ensure that they are done using reliable and appropriate information. The great challenge of fisheries management is to choose and implement the best management strategies to achieve the objectives, despite the fact that there will always be gaps and uncertainties in the knowledge required for fully-informed decisions and actions.

Table 1. The steps normally required in determining an appropriate management strategy to achieve specified operational objectives

Step	Scope	Role of Scientific Information
1. Determine fisheries policy	Applies to whole fisheries sector	Guided by broad information on types of fisheries, nature of resources and ecological context, social and economic characteristics and importance.
2. Set goals	Applies to specific fishery (e.g. as defined by target resources)	<ul><li>Draws on historical performance, including yields, economic performance and social contribution.</li><li>Considers existing problems and opportunities.</li><li>Constrained by scientifically estimated limits.</li><li>May be assisted by formal decision-making techniques.</li></ul>
3. Determine operational objectives and set reference points	Applies to specific fishery. Social and economic objectives may also differ according to sub-sector in fishery (e.g. large-scale commercial, small-scale commercial, subsistence etc.)	<ul><li>Analyses and models used to test, refine and quantify objectives.</li><li>Conflicts between different objectives resolved.</li><li>Target and/or limit reference points defined.</li><li>Requires iterative consultation between decision-makers and scientists.</li><li>May be assisted by formal decision-making techniques.</li></ul>
4. Determine management strategy	Composed of management measures, some of which may be sub-sector specific (e.g. gear restrictions, fishing areas) while others (e.g. closed seasons and areas) may apply to fishery as a whole	<ul><li>Uses analyses, models, and expert knowledge of interested parties to test performance of management measures against operational objectives.</li><li>Determines suite of management measures best able to achieve operational objectives.</li><li>Considers realities of fishing operations in sub-sectors.</li><li>Considers compliance and enforcement.</li><li>Requires iterative consultation between decision-makers and scientists.</li></ul>

2. WHAT DATA AND INFORMATION DO I NEED?

2.1 What information is needed to help make a decision?

In many fisheries agencies, insufficient attention is given to the collection of data and information, and the attempts by these agencies to manage their fisheries are therefore flawed from the outset. Some other agencies go to considerable trouble and expense to collect information on their fisheries, but then do not process and store the information correctly and do not analyse it properly, or at all. Collection of fisheries data is not an end in itself, data stored in log books or on data collection sheets and collecting dust in a cupboard represents a wasted resource. For responsible fisheries management to occur, the required data must be collected and used to obtain information to assist in managing the fishery effectively and hence improving the long-term benefits derived from it. Table 2 summarises the data that are typically required for management. These requirements are determined by the issues and operational objectives the manager needs to consider. They fall into the categories used in Table 2: biological, ecological, economic, social and institutional and can be summarised by the following simple questions that should be continually in the mind of the manager.

– Are current catches in the fishery sustainable and making good use of the resource?

– Are current fishing practices avoiding any damaging and irreversible impact on non-target species in the ecosystem?

– Are the current fishing activities having minimum practical impact on the physical habitat?

- Are other non-fishing activities in the fishing grounds and in the supporting ecosystem being adequately managed to avoid damage and irreversible impact on the ecosystem, including the critical habitats?
- Is the fishery being conducted in an economically responsible and efficient manner consistent with the economic goals and priorities of the country or local area?
- Are those dependent on the fishery for income and livelihoods receiving appropriate, beneficial returns from their fishery-related activities?

If the answer to any of these questions is "No", then the manager needs to consider how the management strategy can be adjusted, through changing management measures, to correct the situation without unacceptable negative impacts on the answers to the other questions. If the manager is unable to answer any of these questions then he or she is inadequately informed to be able to fulfil the mandate of the job properly.

Table 2. Some basic data requirements for providing information to fisheries managers and decision-makers to assist them in selecting suitable management strategies. Additional information on these can be found in the Technical Guidelines (FAO, 1997)

Objective(s)	Data Requirements
Biological	Total landings by major species per fleet per year
	Total effort by fleet per year
	Length and/or age composition of landings for major species
	Discards of major species per fleet per year
	Length and/or age composition of discards per species per fleet per year
	Areas fished by each fleet
Ecological	Total catches of bycatch species (including discarded species), or selected indicator species, per fleet per year
	Length and/or age composition of catches of bycatch species or selected indicator species
	Impact of fishing gear and activities on the physical habitat
	Changes in critical habitats brought about by non-fishing activities
Economic	Average income per fishing unit per year for all fleets
	Costs per fishing unit per year
	Profitability of each fleet (in the absence of detailed economic data this could be based on interviews or similar information)
	Destination of landings from each fleet, and a measure of the dependence on the fishery of other sectors of the community (e.g. processors, wholesalers etc)
Social	Total number of fishers employed within each fleet
	Total number of people employed in fishing or shore-based activities per fleet, by gender and age group where appropriate
	Dependence of fishers and shore-based workers for their livelihoods for each fleet

It is not enough in modern fisheries to attempt to answer these questions with "gut feel" or the unsubstantiated opinions of others. It should be possible for the manager to answer all of these questions making use of good, accurate and recent information, including verifiable information from the interested parties, that will enable him or her to justify the answer and demonstrate the rationale for the answer. The data collected from the fishery will usually be the major source of the information available to the manager. However, these data need to be properly collated and analysed in order to extract meaningful and relevant information for the manager. In most agencies, the scientific division or unit of the agency would be responsible for processing and analysing the basic scientific data.

2.2 Where do I get the information and how can I use it?

Recognising both the importance and difficulties of using good knowledge, the Code of Conduct (Paragraph 6.4) requires that "Conservation and management decisions for fisheries should be based on the best scientific evidence available, also taking into account traditional knowledge of the resources and their habitat, as well as relevant environmental, economic and social factors". This requirement involves three steps:

- the collection of suitable data and information on the fisheries, including on the resources and on the environmental, economic and social factors;
- appropriate analysis of these data and information so that they may be used to address the decisions that need to be made by the fisheries managers, and
- the consideration and use of the analysed data and information in actually making the decisions.

The first of these topics, data collection and monitoring, is a vast topic in its own right and has been the subject of many publications (e.g. see FAO, 1999a) and is also discussed in some detail in the Technical Guidelines to the Code of Conduct: Fisheries Management (FAO, 1997). It is essential that the management agency makes best use of the human and financial resources available to ensure that the most appropriate information for management of the fishery is being continuously collected, is accurate, and is processed and stored in a way in which it can easily be accessed and used.

The second topic, incorporating statistical, stock assessment and social and economic analyses, has been discussed even more thoroughly in the literature. The data collected has to be analysed and processed, typically the function of fishery scientists, economists and social scientists, in order to examine the features and performance characteristics of the fishery which are of interest to the decision-makers. Again, this handbook does not go into the details of the methods of stock assessment and bioeconomic analyses, both of which are disciplines in their own right and have been the topic of many high quality publications. For assistance in these topics, the reader is referred to, for example, Sparre and Venema (1992), Hilborn and Walters (1992) and Seijo, Defeo and Salas (1998), all of which are listed under Recommended Reading at the end of the Chapter. This chapter does consider how the results of these analyses can and should be used, and the third step in the process, the use of the results obtained from the analyses to inform decision-makers, is an important part of the chapter.

In recent years, there has been an increasing awareness of the value of the knowledge and insights of the users of the resource, including traditional knowledge. This is recognised in the Code of Conduct in Paragraphs 6.4 and 12.12, where it is stated that "States should investigate and document traditional fisheries knowledge and technologies, in particular those applied to small-scale fisheries, in order to assess their application to sustainable fisheries conservation, management and development." This is discussed further in this chapter and in Chapter 7.

3. HOW MUCH FISH SHOULD BE CAUGHT: HARVESTING STRATEGIES AND REFERENCE POINTS?

In Chapter 1, Section 7, a management strategy is described as the overall set of measures put in place by the fisheries management authority to regulate the fishery. These measures can include: technical measures relating to gear which are frequently long-term and only occasionally adjusted; technical measures relating to closed areas and seasons which can be put in place over a range of time scales; and input controls, output controls or both, which will often be adjusted more frequently, often annually. The input and output controls are usually central to a management strategy and the actual control value, e.g. the total allowable catch or the permitted effort in a year needs to be determined with care so as to optimise the benefits from the resource in a sustainable way. In setting the control, account must be taken of the status and productivity of the resources, the objectives for the fishery, the needs of the interest groups and the fishing practices in use.

The importance of goals and objectives was emphasised in the Introduction to this Handbook (Section 6 of Chapter 1). As discussed there, the goals for fisheries are frequently expressed in broad terms which are typically too vague to be particularly useful to the manager as actual targets for a management strategy and frequently have conflicts between their different requirements. They therefore need to be developed further if they are to be useful in devising appropriate management strategies, and must be translated into operational objectives. The operational objectives should always be the frame of reference for the manager, both to evaluate how well the management strategy is working and also to evaluate how well the management agency is performing. The manager and decision-makers should regularly be reviewing the management strategy and adjusting it as necessary to ensure that it is the best approach to achieving the objectives. Therefore, the operational objectives must be:

- measurable;
- realistic and achievable;
- accepted by the interested parties in the fishery (Chapter 7), and
- linked to a time-frame.

For example, within a particular fishery, it may have been found that the broad goals presented as an example in Chapter 1 could be best achieved through the following operational objectives:

- to maintain the stock at all times above 50% of its mean unexploited level (biological);
- to maintain all non-target, associated and dependent species above 50% of their mean biomass levels in the absence of fishing activities (ecological);
- to stabilise net income per fisher at a level above the national minimum desired income (economic); and
- to include as many of the existing participants in the fishery as is possible given the biological, ecological and economic objectives listed above.

In this form, these operational objectives are much more specific than the goals and, if the information has been reasonably accurate and the decision-making sound, they will have been selected so that they are simultaneously achievable if a suitable management strategy is developed i.e. there should be no irreconciled conflicts between them. In the hypothetical example, in order to address a potential conflict between maintaining net income per fisher and maintaining employment, it was agreed that net income for the fishers must be maintained above what is referred to as the national desired minimum, but that this may require reducing the number of fishers. Therefore it was (hypothetically) decided to set no minimum limit on employment, but that it would be kept as high as possible without reducing the income below the threshold of those allowed access to fishing.

These operational objectives include reference points such as maintaining the stock above 50% of the unexploited level and the national minimum desired income. Reference points are usually used to guide the manager in setting and adjusting the management measures and provide a guideline as to either a desirable state of the resource or the fishery (a target reference point) or a state to be avoided (a limit reference point). In the example above, both 50% of the unexploited level and the national minimum desired income are used as limit reference points. While similar reference points are used in many fisheries (e.g. $F_{0.1}$ or B_{MSY}), the actual value of any given reference point will be particular to a given resource and fishery, and will need to be estimated for each case and be reviewed periodically.

The target and limit reference points provide signposts for the manager: 'here you are doing well' (target) or 'go any further down this route and we are in trouble' (limit). The manager also needs to know the status of the resource and fishery in relation to those signposts and this requires on-going monitoring of both. The specific characteristics to be monitored are known as performance indicators (or criteria) and relate directly to the reference points: the reference points are specific values of the performance indicators. For example, the performance indicator for the limit reference point of 50% of unexploited biomass would be the current or forecast biomass expressed as a percentage of the unexploited biomass. Net income per fisher would be the performance indicator for comparison with the national minimum desired income.

Once operational objectives, their associated reference points and the performance indicators have been agreed upon, a management strategy that will achieve these objectives must be developed (Table 1). Identifying and selecting the best management strategy requires adequate and appropriate scientific data and information covering all objectives. In practice, developing operational objectives and the management strategy that will achieve them are frequently undertaken simultaneously and interactively, as they are so closely inter-related and require similar information and methods.

3.1 Basic harvesting strategies

Input and output controls are usually set on the basis of one of three basic harvesting strategies (not to be confused with management strategies: a harvesting strategy is one component of the management strategy). The three basic harvesting strategies are: constant catch; constant proportion or constant harvest rate (equivalent to constant effort if catchability of the resource remains the same); and constant escapement (Figure 1). A constant catch strategy will, by definition, result in no change in catch from year to year. However, for the manager to implement a constant catch strategy, that catch must be set low enough to apply in bad years as well as in good years, without damaging the future productivity of the stock, and must therefore be set at a relatively low level. Therefore the fisher pays a price for the absence of inter-annual variability in catch in a constant catch strategy by foregoing potential catch in good years. In a constant proportion strategy, the effort remains constant and therefore there will be changes in catch from year to year as the resource varies over good, bad and intermediate years. This variability results in some uncertainty about future catches for the fisher compared to the constant catch strategy. It also has benefits for the fisher, though, as it means the catches will be higher in good years, in contrast to the constant catch strategy, generally leading to a higher annual average catch. A constant escapement strategy (or constant stock size strategy) would aim to ensure that a constant biomass, sufficient to maintain recruitment, was left at the end of every fishing season. This type of strategy tends to achieve the highest annual average catches of the three categories but with the highest variability, in many cases including zero catches in some years.

The decision on which type of harvesting strategy to pursue should be made from a knowledge of the requirements of the fishery and with consultation with the interest groups on the trade offs they would like to make between maximising catch and minimising variability. The much more difficult question is, given one of the strategies, how does the manager decide on the actual catch, effort or escapement which should be set under the strategy. This is discussed in later sections of the chapter. It should also be noted that these harvesting strategies could all be pursued through the use of output control (setting a TAC), input control (setting the effort that can be expended in a year), or even the use of closed seasons (which can be a form of output control – see Chapters 3 and 4).

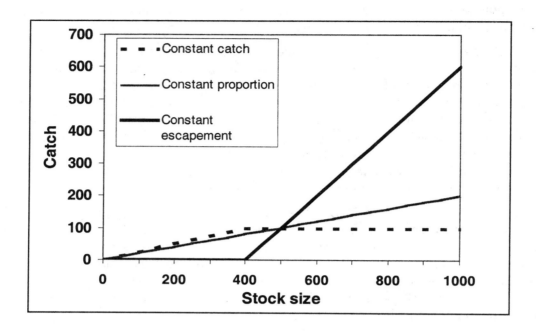

Figure 1. Simple examples of the three classes of harvesting strategy and their relationship to stock size: constant catch (with provision for a linearly decreasing catch when the stock size falls below 400); constant proportion; and constant escapement (after Hilborn and Walters, 1992).

3.2 The classic reference point : maximum sustainable yield

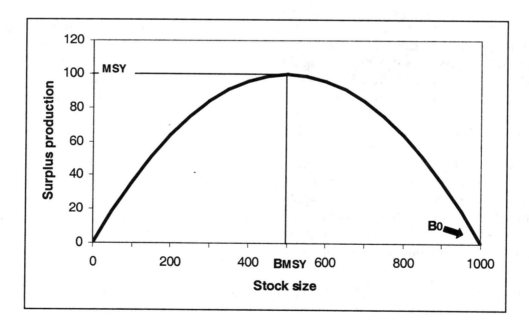

Figure 2. Schaefer model of surplus production (biomass dynamic) as a function of stock size showing the major reference points. Other forms of surplus production model can have BMSY at a higher or lower stock size than the 50% of B0 of the Schaefer model. MSY = maximum sustainable yield; BMSY = the biomass at which MSY occurs; and B0 = the average unexploited biomass of the stock (the average 'carrying capacity').

In the 1960s and 1970s, maximum sustainable yield (MSY) was seen as the ideal target to aim for in managing fisheries, and managers attempted to obtain MSY through striving to set the MSY as a target catch level or to determine the fishing mortality rate that would generate MSY (F_{MSY}). The maximum sustainable yield concept is based on a model, referred to as a surplus production or biomass dynamic model (Figure 2), which assumes that the annual net growth in abundance and biomass of a stock increases as the biomass of the stock increases, until a certain biomass is reached at which this net growth, or surplus production, reaches a maximum (the MSY). This biomass is referred to as B_{MSY}, and the fishing mortality rate which will achieve MSY is similarly referred to as F_{MSY}. As the biomass increases above B_{MSY}, density dependent factors such as competition for food and cannibalism on smaller individuals start to reduce the net population growth which therefore decreases until at some point, the average carrying capacity of the stock, net population growth reaches zero. In reality, an unexploited stock will tend fluctuate about this biomass because of environmental variability.

MSY was such a well established target for managing fisheries that it is included in the 1982 United Nations Convention on the Law of the Sea (LOS), where it is stated that coastal management agencies should "... maintain or restore populations of harvested species at levels which can produce the maximum sustainable yield, as qualified by relevant environmental and economic factors".

This requirement of the LOS is equivalent to specifying a limit reference point of B_{MSY}. This is not the same as setting MSY as a target reference point for catch, however, and using MSY as a target reference point has been found to be dangerous. This is because it is impossible to estimate MSY precisely for any stock. If MSY is over-estimated, then a fishery will be allowed to take more than the maximum production of the stock which will cause a reduction in the biomass every year. In a new fishery this could drive the biomass down to the level at which MSY is produced (B_{MSY}) but if continued after that will drive the biomass down further, where annual production gets smaller and smaller, making the situation even worse. Even if average MSY could be precisely determined, the productivity of a stock varies from year to year under the influence of environmental variability. Therefore if the stock is at B_{MSY}, in some years production may still be less than MSY and, if MSY is taken as the catch, the biomass will be driven below B_{MSY}, possibly driving the stock into a downward spiral. Therefore MSY is no longer seen as a target reference point for fisheries managers to strive for, although it can still be used as a limit reference point i.e. as an upper limit to the annual catch, which should be avoided.

3.3 Reference points based on fishing mortality rate

A standard assumption in stock assessment is that:

Catch = (Fishing effort) X (Catchability per unit of effort) X (Abundance of the stock)

From this, it can be seen that if catchability remains constant each year, then the fishing mortality rate (catch as a fraction of abundance) is directly related to effort: the higher the effort the higher the fishing mortality rate i.e.:

Catch/(Abundance of the stock) = (Fishing effort) X A constant

Therefore, a strategy which attempts to maintain a specific fishing mortality rate is equal to a constant effort strategy as long as catchability remains constant. A desirable (target) fishing mortality rate should be determined from examining the productivity of the stock (through a stock assessment) and could be based on, for example, yield and biomass-per-recruit considerations or, as with MSY, on surplus production considerations.

Yield and biomass-per-recruit methods examine the individual growth and mortality rates of a species or stock and use these to model the proportion of each recruit (perhaps easier to think of as a percentage of 100 recruits) that would be caught by a fishery at a given fishing mortality rate. As with the surplus production model, it is usually found that this yield-per-recruit increases with increasing fishing mortality (or effort) up to a maximum, and then begins to fall as effort increases above that maximum (Figure 3). The fishing effort which results in the maximum yield-per-recruit is referred to as F_{max} and could be used as a target reference point. However, before selecting this as a reference point, the affect of F_{max} on the spawner biomass should be checked. As shown in Figure 3, surviving spawner biomass falls continually as F is increased and one needs to select a fishing mortality rate that not only achieves a good yield-per-recruit but also leaves a sufficiently high spawner biomass (as indicated by spawner biomass-per-recruit) in the water to ensure good recruitment is maintained. It is commonly accepted that for many stocks the minimum desirable spawner biomass is between 30 and 40% of the spawner biomass in the absence of fishing. Figure 3 shows the fishing mortality that would result in spawner biomass being reduced to 40% (referred to as F_{SB40}). In this example, F_{SB40} is considerably lower than F_{max}, so some trade off in short-term maximum yield would be required in order to ensure spawner biomass is kept high enough to ensure sustained recruitment.

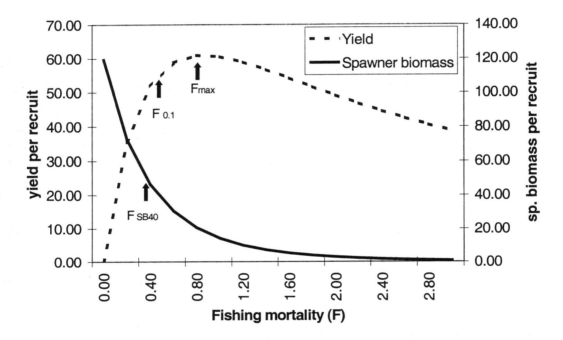

Figure 3. Yield and spawner biomass-per-recruit plots for a hypothetical snapper stock, showing common reference points for fishing mortality: F_{max}; $F_{0.1}$; and $F_{SB40\%}$. See text for explanation of the different reference points.

The third reference point shown in Figure 3 is $F_{0.1}$, which is widely applied as a target reference point. Although the definition of $F_{0.1}$ may seem confusing, it is relatively easily calculated from a yield-per-recruit curve using, for example, a spreadsheet and a minimisation routine (e.g. Solver in Excel). $F_{0.1}$ is defined as the F value at the point where the slope on the yield-per-recruit curve is 10% (0.1) of the slope at the point where F is 0 (the initial slope). There is no theoretical rationale for the use of the $F_{0.1}$ reference point except that it will always be less than F_{max} and hence result in a higher spawner biomass after fishing (e.g. Figure 3) and it has been found, in general, to be quite robust to important uncertainties. In the example shown, in order to choose between an $F_{0.1}$ and a F_{SB40} strategy, the scientists and decision-makers would need to consider the accuracy of their per-recruit data and results, and aspects of the biology of the species such as variability in average recruitment and the natural mortality rate of the species.

3.4 Reference points based on size-limits

In many fisheries there is little information available on the biomass of the stock and estimates of F and M, even if available, may be very unreliable. This frequently applies in small-scale fisheries, especially (but not exclusively) in developing countries. In such cases, a minimum precautionary approach could be to ensure that no immature fish are caught in the fishery and that a reasonable proportion of the fish in the stock have the opportunity to reproduce. This requires specifying the minimum size of fish (typically expressed as a length measurement) which can be caught. The minimum size would then be a limit reference point for the fishery. Clearly this could only be considered when the fishing gear or method being used is sufficiently selective for fishers to be able to target specific size ranges and when the regulation can be enforced. Size-based reference points would typically be implemented through gear restrictions (Chapter 2), possibly complemented by area or time closures (Chapter 3). An appropriate minimum size to be set as a reference point could be identified by looking at the relationship between the size of the species and the percentage which have reached maturity or by looking at spawner biomass-per-recruit curves with different ages at first capture. Size-based reference points to set limits on minimum size at first capture have been widely applied in invertebrate fisheries and, with suitable minimum sizes and good enforcement have frequently been successful. As with all management measures, however, they would normally not be adequate as the only measure and would be implemented in combination with others in order to address the full range of objectives.

3.5 Multi-species and ecosystem-based reference points

The reference points discussed above are all single-species reference points and assume that only one species is being fished for and managed. In practice very few fisheries are truly single-species and range from fisheries with a small bycatch of other species to those with a wide diversity of species in the catch, perhaps without any single-species being dominant. The Code of Conduct requires that "... catch of non-target species, both fish and non-fish species, and impacts on associated or dependent species are minimised" and also that "biodiversity of aquatic habitats and ecosystems is conserved and endangered species are protected" (Paragraph 7.2.2). These stipulations require that multi-species and ecosystem impacts are also taken into account when determining management strategies and should therefore be considered at the same time as selecting appropriate reference points to guide management of the fishery.

In order to achieve responsible ecosystem-based fishery management, the manager should identify and apply ecosystem reference points and then set a management strategy in accordance with those reference points. However, genuine ecosystem-based reference points are rarely, if ever, used and the best approach at present is usually to develop a suite of single-species or single-factor indices and to manage according to those. Under this approach, not only would reference points be developed and applied for the major target species, but also for key by-catch species, indicator species and species identified as being vulnerable or depleted. The strategy would be developed based on the full suite of reference points and make use of gear regulations, closed areas and/or closed seasons to minimise undesirable impacts on non-target species. In addition, if sustainable catch of a vulnerable by-catch species required a lower amount of effort in a fishery than the desirable target effort for the main target species, the effort should be set at the lower level to ensure sustainable fishing on the vulnerable species. In practice, this approach could well lead to a need for more conservative fishing, and hence a short-term reduction in social and economic benefits. In the longer-term, however, such an ecosystem-based approach applied to many existing fisheries should lead to an increase in the quality and hence value of the catch through allowing valuable depleted species to recover from decades of over-exploitation. In many cases, where over-exploitation has been severe and sustained, it could also lead to an increase in quantity as the many over-exploited species recover to more productive biomasses.

Alternative, or additional, indicators could be, for example, monitoring over time the percentage contribution of species to catches, tracking indices of species diversity in the catches or population, and monitoring length frequencies of indicator species or taxa to check for signs of growth over-fishing.

3.6 Economic and social reference points

Other reference points consider the economic performance of the fishery and include maximum economic yield (MEY) and the fishing mortality rate (F_{MEY}) and effort (f_{MEY}) which achieve maximum economic yield. They can be estimated from, for example, the Gordon-Schaefer model (Figure 4) which combines a surplus production curve for the stock in question with the relationship between the cost of fishing and fishing effort. The Gordon-Schaefer model can also be used to estimate the theoretical bioeconomic equilibrium (BE) in an open access fishery, at which point costs and revenues are equal so that there is no incentive for new entrants to join the fishery (Figure 4).

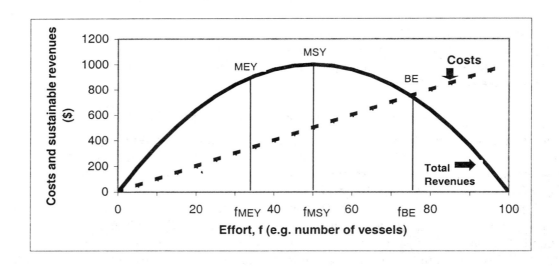

Figure 4 Gordon-Schaefer bioeconomic model of costs and sustainable revenues for a fishery as a function of fishing effort. MEY = maximum economic yield, MSY = maximum sustainable yield, BE = the bioeconomic equilibrium (see text for explanation). The suffix f indicates the effort at each of those reference points.

Reference points can also be set on the basis of other economic or social performance indicators and should be established from the operational objectives of the fishery and consideration of the monitoring capacity of the management agency and fishing groups. They could include indices of employment, income per person or fishing unit, age composition of the fishers, levels of satisfaction or any other measure of the benefits, or opposite, being generated by the fishery or fisheries.

4. WHAT TOOLS CAN BE USED TO GENERATE INFORMATION TO ADVISE MANAGEMENT?

In the case of, for example, a pharmaceutical company trying to develop a new drug, the best medication (management strategy) to cure an illness (the operational objective) will be determined by undertaking an intensive series of laboratory tests. The results of these tests will inform the company as to whether any of the drugs they have been developing will provide a cure and should be commercially produced. Unfortunately, controlled tests of this nature are not possible in fisheries management but, wherever enough data on the fish and fishery are available, mathematical calculations and projections, ranging from relatively simple to very complex, can be used as a type of laboratory and thereby to advise the decision-makers on the status of the fisheries and what, if any, adjustments are needed in the existing management strategies. A primary purpose of the data collected by the management agency is for these purposes.

A message emphasised in this Handbook is that fisheries science is still an imprecise science and there are limits to our knowledge of the dynamics and behaviour of individual stocks and even more so of communities and ecosystems. In many cases what we don't know far outweighs what we do! Nevertheless, by monitoring trends in populations and communities, by observing their responses to fishing and to environmental factors, we can gain invaluable information on how they are likely to respond in the future, including to changes in management strategy. In keeping with most aspects of human knowledge, the closer the forecast situation is to circumstances that have been experienced before, the more reliable it is likely to be. Put another way, beware of long-term forecasts and forecasts that go far beyond previously experienced conditions.

Just as laboratory tests can be good and bad, so can mathematical tests and models. The scientific staff of the management agency are responsible for trying to develop the best mathematical methods with the resources and data that they have available in order to:

- provide the information required by the decision-makers;
- be sufficiently accurate to minimise the chances of making incorrect decisions; and
- reduce the uncertainty remaining in the answers to a low enough level for the decision-makers to be reasonably confident that their selected strategy will work.

They must also ensure that the decision-makers are aware of uncertainties and potential errors in the estimates and forecasts. Fisheries managers and interested parties are generally interested in the net production of a resource and how much of that can be taken by a fishery. Net production is composed of three basic processes: recruitment of new individuals to a population through reproduction; the sum of the individual growth of all the members of a population; and the total mortality, which can be divided into the individuals caught and killed or removed by the fishery (fishing mortality) and the members killed or dying by any other cause (natural mortality). All stock assessment methods attempt to determine those rates directly or indirectly, and to consider how they could change at different population sizes, under different management strategies and, where considered, under different environmental and ecological conditions.

4.1 Single-species methods

Single-species methods of stock assessment, as their name implies, consider only the population or stock of a single-species or species-group at a time and generally make the assumption that the dynamics of the population (recruitment, growth, mortality) are affected only by the abundance or biomass of that stock and the affect of fishing on it. This assumption obviously ignores the effects of the environment and of other populations, such as the abundance of predators and prey, on the stock. The reasons why single-species methods make these blatantly incorrect assumptions is because there is an underlying population effect which is important to understand in managing fisheries and because the interactions between environment, community and the stock of interest are frequently so complex and so poorly understood that it is often impossible to build models that reflect any verifiable understanding of this reality. The underlying population affect is sufficiently important that in most cases where good single-species assessments are undertaken using reliable data, they do provide invaluable information for the management of that stock. Despite their limitations, they should therefore not be discarded but every effort should be made to ensure that the method is appropriate for the resource and the questions being asked, and that the data are as reliable and complete as practically possible. Results from these models should also be supplemented by information on the fishery and ecosystem from other sources, including the interested parties, socio-economic studies and the use of ecosystem indicators and models.

Table 3. Main categories of single-species stock assessment methods and their characteristics.

METHOD	MAIN INFORMATION REQUIRED	COMMENTS
A. Production models	-Annual catch -Annual index of abundance e.g. cpue or biomass estimate	- Do not consider age structure of catch or population - Estimate parameters and variables such as MSY, effort at MSY, mean unexploited stock size, biomass time series etc. - Very widely applied e.g. tuna commissions, south east Atlantic - Caution should be used, especially when fitting with equilibrium methods - Good estimates require good data contrast in effort and biomass
B. Size and age-based models		
B1. Yield and biomass-per-recruit	-Somatic growth rate -Natural mortality rate -Age/size at recruitment to fishery - Selectivity of gear for different age/size classes -Mean size at sexual maturity	- The Beverton and Holt per-recruit models assume knife-edge selectivity and constant fishing mortality and natural mortality for all ages. The general models avoiding these assumptions are preferred. - Assume the stock is in equilibrium i.e. that the biomass and age –structure are constant from year to year. - Assume that recruitment is constant from year to year, which is likely to be false at high fishing mortalities when low spawning biomass may reduce recruitment.
B2. Virtual population analysis (VPA) and cohort analysis	- Number of fish caught per age class.	- One of most powerful assessment methods available. - Provides estimates of past stock abundances, size-selectivity in fishery and estimates of recruitment to fishery. - Requires independent estimate of F for cohorts still present in the fishery (terminal F's), either from assumption or by direct estimate from surveys or mark-recapture. - Assumptions on terminal F's and M are probably the greatest source of error in VPAs.
C. Stock recruit models	- Separate estimates of stock and recruitment over a number of years	- Recruitment will almost certainly drop if the stock size is reduced sufficiently and managers must take this into account. - Stock size is only one determinant of recruitment, and recruitment will vary substantially around the mean stock-recruit relationship i.e. uncertainty in forecast recruitment will be high even when a good relationship has been determined.

Single-species methods have been intensively studied and applied for decades and many different approaches now exist for different circumstances and different fish types. These are summarised in Table 3. While there are different ways of categorising the methods, in this table they are listed under three categories: surplus production or biomass dynamic models, size/age-based models and stock:recruit models. Each has different data requirements, makes different assumptions and enables different questions and scenarios to be addressed. This is summarised in the table. None of the stock assessment methods are perfect and, as discussed in Section 9, the results of all will be affected by process, observation and estimation uncertainty. The manager must be informed of this uncertainty and how it could affect the results of the stock assessment. The known uncertainties must be considered when choosing management measures and strategies, by considering the potential errors in assessment and decision-making and by choosing options which are robust to the more likely errors. This is an important example of where intelligent application of the precautionary approach is essential.

4.2 Multi-species methods

Single stocks and populations can be affected by other species in the ecosystem in two ways, through biological interactions and technological interactions. Direct biological interactions occur when a species is a predator, prey or competitor of another, in which case any change in abundance and distribution of either species will affect the dynamics of the other. These effects are ignored in single-species models. Indirect biological interactions also occur, for example when a third species, not directly interacting with the first, is affected by changes in the abundance of the first through their impact on a second species directly interacting with both.

Technological interactions occur when one species is affected by the fishing activity on another species, for example if it is caught as a bycatch, whether landed or discarded. In general technological interactions are easier to quantify and hence consider in fisheries management than are biological interactions which are more complex and dynamic. Multi-species per-recruit models are particularly useful for consideration of technical interactions and not especially demanding of data and expertise.

The sum of all these interactions leads to a fundamental principle of fisheries management: that the yield from a multi-species fishery will always be less than the sum of the potential sustainable yields of all the separate species (see Table 1, Chapter 1). Recognising this principle, fisheries managers should be supplementing the information from single-species approaches with that from multi-species and ecosystem tools, allowing them to consider the implications of this principle on their overall objectives and to identify strategies to optimise the yields that realistically can be obtained from the ecosystem as a whole.

It is therefore being increasingly recognised that fisheries management must move from seeing fisheries as dealing with single-species to considering fisheries as the multi-species, ecosystem-based activities that they invariably are. However, the amount of uncertainty is generally much higher as one attempts to include more factors in a problem and this certainly applies in fisheries. Further, as one considers more and more issues and objectives, which is an inevitable consequence of considering the whole ecosystem, the number of potential conflicts and constraints increases dramatically. For these reasons, the necessary move from single-species to ecosystem-based management has barely started in most countries and fisheries. Nevertheless, there are some important tools available for considering these interactions and they should be used to help inform managers in making decisions, with the same careful and precautionary consideration of likely uncertainties and errors as discussed for single-species methods.

The better developed and more commonly applied approaches are shown in Table 4. Of these, multi-species virtual population analysis will be too demanding of data for application in most fisheries and multi-species surplus production approaches are likely to be equally impractical. Aggregated production models, multi-species per-recruit models and dynamic trophic level models have all been applied and found to provide information of relevance and use to fisheries management objectives and strategies.

Table 4. Main categories of multi-species and ecosystem-based assessment methods and their characteristics

METHOD	MAIN INFORMATION REQUIRED	COMMENTS
A. Multi-species extensions of single-species methods		
A1. Multi-species surplus production models	Same as for single-species +indices of abundance +, preferably, abundances of all species with important interactions with the 'dependent' species.	In theory, enable consideration of biological interactions, but of little practical value because: - if only indices of abundance are available for the species included, then enormous statistical problems will be encountered in estimating the parameters; and - as for single-species, good data contrast is required for good estimates.
A2. Aggregated production models	- Annual catch aggregated into appropriate species groups -Annual index of abundance e.g. cpue or biomass estimate for same aggregated groups	-Has proven informative in some cases where tried -Provides a feasible source of information for ecosystems with high species diversity -Caution required as the selected reference point for the aggregation could lead to depletion of some vulnerable species while producing sustainable yield for the aggregation as a whole
A3. Multi-species per-recruit models	- As for single-species per-recruit analyses. - The relative catchability of each species for a unit of fishing effort - The relative recruitments of the different species	- Can be used for more than one fishery at a time as well as more than one species - Consider technical interactions, not biological interactions - Involve the same assumptions and limitations as single-species per-recruit methods - A useful tool for assisting in setting reference points in multi-species fisheries
A4.. Multi-species stock recruit models	- As for single-species method - Abundance estimates of other predators and competitors on the species of interest.	- Extends single-species stock recruit models to consider the effect of other species on a given species
B. Multi-species VPA	- As for single-species method - Estimates of the number at age of individuals of the species of interest consumed by all other species	- Has the potential to provide very useful information taking into account some biological interactions - Very data intensive and therefore probably not applicable in most circumstances
C. Food web and trophic level models	- Estimates of biomass of all major species or species groups - Production per unit biomass for each group - Consumption per unit biomass per group - Average diet composition per species group	- The requirements listed here are for a simple food web type model, models incorporating e.g. physical factors require more - In equilibrium form useful for gaining insight into trophic relationships and direct and indirect interactions - In dynamic form (e.g. Ecopath with Ecosim) can be used to explore multi-species implications of harvest policies - Invariably include substantial uncertainty which must be rigorously explored, reported and considered

4.3 Considering the benefits to society

Fisheries exist to provide benefits to humans, and fisheries management should be attempting to optimise those benefits within its objectives. The single-species and multi-species methods described in the previous two sections were focused on the resources, and the benefits in the methods described there are reduced to a single measure: the catch or yield. In a simple single-species, single fleet fishery where all the interested parties have the same objectives, this may be an adequate measure of benefits. Nevertheless, even then the costs of fishing, and hence the net profits, are likely to change with stock abundance, and fishing effort should be explicitly considered. In more complex cases, such as where different fishing groups using different methods are exploiting the same resources or where multi-species resources are being exploited, maximising the yield on a sustainable basis is unlikely to be a full or adequate socio-economic objective, and information on the economic and social benefits for the different groups that can be expected under different management strategies will also be important for decision-making.

Just as there have been many stock assessment tools developed to deal with different types of data and different questions, so there have been many different models developed to extend such assessments to include, particularly, economic performance. A commonly applied bioeconomic model is the Gordon-Schaefer model which, as its name implies, uses the Schaefer surplus production model shown in Figure 2 as the underlying biological model. The Gordon-Schaeffer model was discussed earlier and illustrated in Figure 4.

Age structured bioeconomic models, some including spatial distribution of the resource and fleets have also been developed, allowing the investigation of management strategies that exploit different size or age classes of the resources and in which spatial patterns are significant. Good examples of such models can be found in, for example, Seijo et al. 1998 and Sparre and Willmann 1993, both of which are listed in the Recommended Reading list at the end of this chapter.

In the same way that bioeconomic models can be developed from the standard stock assessment models, so they could be adapted to include social factors. For example, it may be useful to consider effort in the Gordon-Schaefer model in terms of number of fishers, giving a measure of employment as well as revenue. Similarly, if there is a relationship between yield from the fishery and the number of fishing and shore-based jobs, these models can be used to investigate how different management strategies will affect employment. Ecopath with Ecosim can include basic social and economic characteristics of the different fleets fishing on an ecosystem and can therefore be used to investigate the biological, ecological, economic and social impacts of different harvesting strategies within the ecosystem as a whole. This facility has the potential to provide useful information to supplement that obtained from the single-species models, which typically contain more detailed information on that specific resource.

5. HOW IS THE INFORMATION USED TO DEVELOP A MANAGEMENT STRATEGY?

5.1 What sort of biological information is needed?

A primary consideration in selecting a management strategy is the impact of each strategy on the status of the stock, stocks or fish community. In cases where time series of catch and effort information are available, it may be possible to analyse trends in the catch-per-unit-effort, which with careful interpretation may provide an indication of trends in resource abundance. Such trends can indicate when adjustments in the management strategy are required, for example to prevent on-going declines. Catch and effort data may also permit more sophisticated analyses, such as the application of biomass dynamic (surplus production) models and, with additional information on the length or age structure of the catch, virtual population analysis (VPA). More sophisticated analyses allow more sophisticated biological reference points to be determined, enabling the management agency to aim for, for example, obtaining the maximum average sustainable catch, instead of simply aiming to avoid on-going declines. Figure 5 shows the estimated biomass obtained from fitting a biomass dynamic model to catch and effort information from the sardinella fishery in Angola. From this information, the manager could see that the sardinella stock had been recovering from a period of over-exploitation since the catches had declined, and the estimated reference points showed there was scope for increasing the annual catches from their levels in recent years.

Catch and effort data from a fishery are generally the cheapest and easiest information to obtain, and collecting (and using!) good estimates of annual catches and effort should be a fundamental task of the scientific branch of the management agency.

Yield and biomass-per-recruit calculations can also be used to provide information on the status of the stocks, the impact of fishing on them and how to adjust fishing mortality to achieve desired operational objectives through appropriate management strategies. Per-recruit methods require estimates of the growth rate of the species being assessed, their natural mortality rate (which can be approximated from knowledge of the growth rate), and the selectivity of the fishing gear for different size groups or, at least, the size at which the species become vulnerable to capture (Table 3). While this information may appear more difficult to acquire than simple time series of catch and effort, it can be obtained from good time series of the length frequencies of the catches and in some countries these have been collected, even though monitoring of catch and effort has, regrettably, been discontinued.

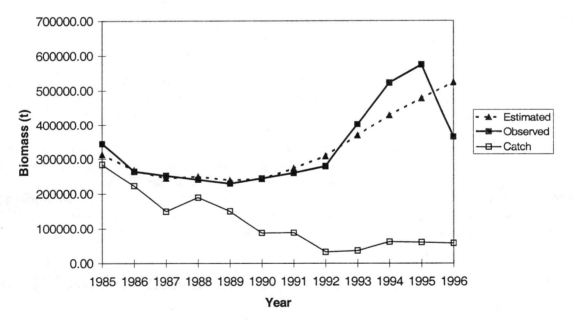

Figure 5. Estimates obtained from fitting a biomass dynamic model to catch and effort data for the fishery for sardinella in Angola. The chart shows the <u>Catch</u>, and the <u>Observed</u> biomass (indicated by catch-per-unit-effort data) compared to the biomass <u>Estimated</u> from a biomass dynamic model.

Consideration of the yield and spawner biomass-per-recruit curves for a particular species and gear type (Figure 3) enables the manager to determine what level of fishing mortality will achieve a good yield-per-recruit while at the same time maintaining a high enough spawner biomass-per-recruit to sustain recruitment. In addition to estimates of appropriate reference points, it is also necessary to consider the current level of fishing mortality in relation to the fishing mortality required to achieve the target. With the same data, an initial estimate of fishing mortality can be obtained through undertaking a catch curve analysis on good estimates or samples of the population length frequency (see e.g. Sparre and Venema, 1992). However, as with all stock assessment methods, accurate and precise estimates are best obtained through the use of time series (minimum 3-5 years, depending on the application and data) of at least good catch and effort information.

The potential yield from a stock is dependent on the average size and age of the fish taken by the fishery and there is generally an optimum average size, below which there is considerable risk of over-exploiting the stock and above which potential yield from the resource is lost. The size-selectivity of the gear used in the fishery is therefore important in managing the fishery, as discussed in Chapter 2. Per-recruit analyses can provide useful information on how changes in selectivity of the gear can influence the potential yield from the resource and the probable survival per-recruit, helping in the selection of appropriate gear. Figure 6 shows the biomass-per-recruit for a western Atlantic snapper caught in two different fleets: as bycatch in an offshore trawl fleet targeting mainly shrimp and in a nylon gill net fishery targeting fish. The former, designed to catch the smaller shrimp species, catches the fish at a smaller size than the gill net, and hence has a considerably greater impact on spawner biomass of snapper at a lower fishing mortality. Per-recruit analyses on the two species indicated that at a fishing effort suitable for sustainable utilisation of shrimp, the snapper and other fish species will be severely over-exploited. Using such approaches, it is possible to consider the trade-offs required, for example, in foregoing yield in the shrimp fishery in order to ensure the sustainable use of the snapper resource.

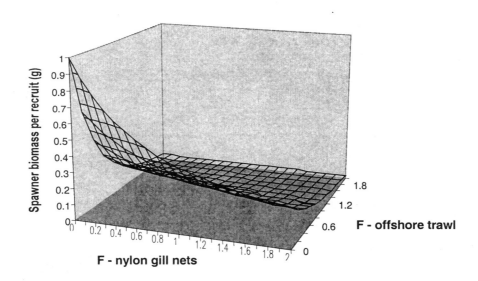

Figure 6. Spawner biomass-per-recruit of a snapper species under different fishing mortalities (F) for two different gear types[1]. The nylon gill nets catch the snapper at a much larger size than the shrimp-directed offshore trawl gear.

All stock assessment approaches require making certain assumptions about the data and the dynamics of the resource. Two common and important assumptions are that catch-per-unit-effort (CPUE) is proportional to the abundance or biomass of the resource, and that natural mortality rate, M, is constant for all ages of fish and in all years. The assumptions behind any stock assessment, in fact any source of information and decision, are important and should be

---

[1] From FAO (1999). Meeting report of the second CFRAMP/FAO/DANIDA stock assessment workshop on the shrimp and groundfish fishery on the Brazil-Guianas shelf. Georgetown, Guyana, 18-29 May 1998, FAO, Rome. 41pp.

considered when using information and when making decisions. The use of per-recruit analyses (and most of the length frequency analyses that are often used to estimate parameters such as growth rates and total mortality rate for per-recruit analyses) assume that recruitment will remain constant. In practice, one of the greatest sources of uncertainty in resource dynamics, and hence in fisheries management, is the very high variability from year to year in recruitment of young animals to a stock, which can vary by an order of magnitude or more from one year to the next (Figure 7). When providing advice on the effect of management measures, the scientists should also consider the impact of variability of recruitment on their results and on the attainment of the operational objectives.

Useful information on the status of the stocks can also be obtained by examining their size structure to determine whether there have been any major changes over time. A significant decrease in the average size (normally indicated by length) of a stock may indicate growth overfishing, suggesting that the larger individuals are being removed at a rate too high for sustainable utilisation. Conversely, the decrease may be the result of good recruitment in recent years. The two different scenarios would require very different responses in management and the scientists need to ensure that they have the data and undertake the analyses necessary to determine the cause of the change. Similarly, an increase in the average size may indicate poor recruitment in recent years resulting in older and larger animals being exploited at proportionately higher fishing mortality rates, or a decrease in effective fishing mortality rate possibly leading to more fish surviving through to attain larger sizes. Again, the underlying cause for this should be investigated and the appropriate management response considered. Where relatively sedentary species are being considered and where closed areas are in existence, comparing the length frequency of the exploited portion of the stock with that of the sub-population in a closed area may also give an indication of the effects of exploitation (see Chapter 3).

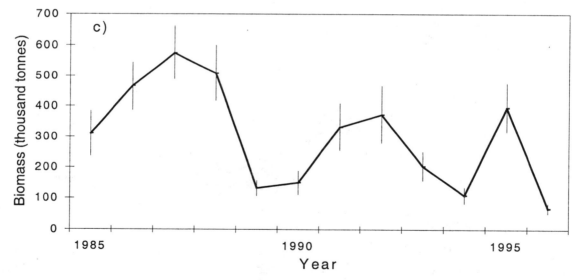

Figure 7. Time series of recruitment biomass estimates in the South African anchovy stock demonstrating high variability in this short-lived species. The vertical lines show one standard deviation of the estimated mean on either side of the estimate, giving an indication of the uncertainty in the estimates. The 95% confidence limits of the estimate would be approximately double the length of each vertical line.

The above examples do not discuss in depth how to take uncertainty into account in the analyses but this topic is introduced in the Appendix to this chapter. Methods for including uncertainty in assessments include the following.

- Sensitivity analyses, for data and assumptions, in which the impact of a change in a parameter or assumption on the output from a model is explored.

- Monte Carlo analyses, in which, instead of undertaking an analysis once with fixed values of all 'known' parameters and variables, the analysis is run a large number of times, each time selecting a different value of the parameters from a pre-specified distribution. These analyses will generate a large number of estimates of the unknown parameters and variables, giving an indication of the range and distribution of possible values for each unknown.

- Bayesian approaches are essentially extensions to existing deterministic methods, as are Monte Carlo approaches, with the important advantage that they can make use, in a formal statistical manner, of other sources of information in addition to the data available for the stock under consideration. For example, Bayesian approaches could use estimates of key parameters from other stocks or expert opinion on possible parameter values in fitting a biomass dynamic model to fisheries data, to help "inform" the estimation procedure of likely values of these parameters in a particular case. They therefore have particular potential value in fisheries where there are only limited data available from the fishery itself.

5.2 What sort of ecological information is needed?

Table 2 showed that including ecological considerations in fisheries management adds to the demands on the data collection and analysis requirements of the responsible agency, increasing the number of species which need to be monitored as well as requiring information on ecosystem interactions and the state of the different habitats occurring in the ecosystem. This could be seen as a luxury but has come to be recognised as being essential to sustainable and efficient use of the resources. The target species are dependent for their survival and productivity on the ecosystem in which they live, and changes in that ecosystem will affect the resources. The manager needs to be aware of any such changes, whether natural or caused by fishing or other human activities. This not only enables an assessment to be made of the likely impacts of the changes, and the management strategy to be adapted accordingly, but also allows for the adoption of management strategies which minimise damage to the ecosystem. Minimising damage will require reference points to be developed for those ecosystem components identified as being of particular importance or particularly useful as indicators of some ecosystem property (FAO, 1999b). The status of the ecosystem can then be monitored against those reference points.

In the case of impacts of a management strategy on non-target, associated and dependent species, the performance indicators are likely to be similar to those for the target species. However, there may be less data and information available for the former with which to estimate the performance of the strategy, and the types of reference points may need to be modified from those typically used for target species, so as to require less precise information. Similarly, the impact of different fishing gears on the physical habitat may be impossible to estimate quantitatively, and gear types or strategies may need to be ranked against such criteria as, for example, good, neutral, or harmful. Table 1 in Chapter 2 provides a useful approach to ranking the ecological impact of different fishing methods. When such rankings can be made, the information should be provided to the decision-makers. In cases where no information is available for setting reference points or evaluating the ecosystem against important reference points, this should be clearly stated by the scientists to ensure that the decision-makers are aware of the uncertainties and lack of information on these issues.

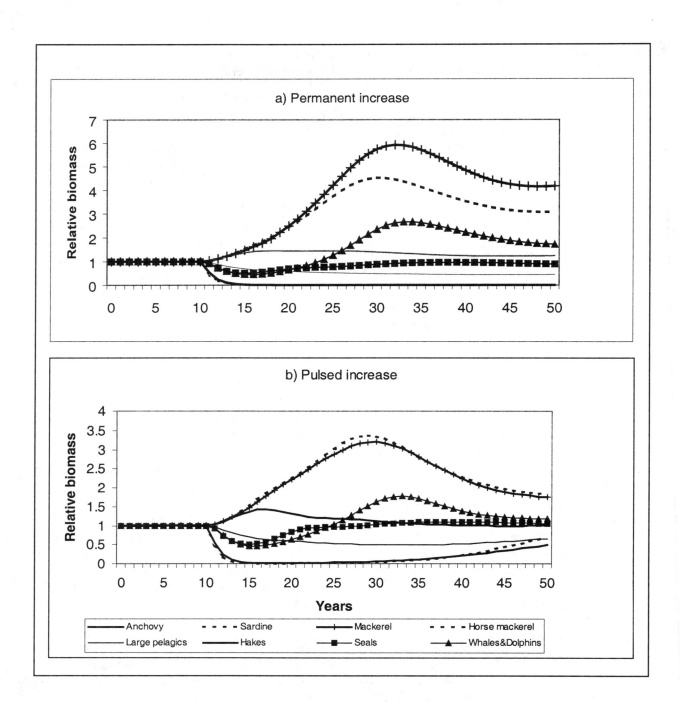

Figure 8. Simulated effects of increased fishing effort in the small pelagic fishery on selected taxonomic groups in the southern Benguela ecosystem. a) Results from a fourfold increase in fishing mortality on small pelagics from year ten onwards and b) from a fourfold increase in F from years 10 to 15 only. Relative biomass is the biomass as a proportion of the biomass at the start of the simulation. Figure modified from Shannon, L.J., Cury, P.M. and Jarre, A., 2000. Modelling effects of fishing in the southern Benguela ecosystem. *ICES Journal of Marine Science* **57**:720-722.

Our knowledge of ecosystem dynamics is notoriously incomplete but suitable models representing our best understanding can still be informative. Figure 8 shows a simulation from an Ecopath with Ecosim[2] model of the southern Benguela ecosystem under two different management strategies, in this case both a simple modification of fishing mortality on the main commercial small pelagic species: sardine, anchovy and roundherring. The simulation estimated that in addition to the target species, a large number of other species will also be affected by the changes in fishing mortality. For example, the biomass of chub mackerel is estimated to increase by up to 6 times its starting biomass while that of large pelagics will decrease by nearly one third of its starting biomass when the increased fishing mortality is maintained (Figure 8a). Scientific understanding of ecosystem interactions and dynamics is still very limited and there is therefore a high degree of uncertainty in any predictions of ecosystem behaviour but scientists should still consider the ecosystem implications of different management strategies, and models can assist in this, as they can in single-species cases.

5.3 What sort of social and economic information is needed?

Fisheries exist to provide social or economic benefits to society, and it is a task of the manager to ensure that these benefits are obtained in an appropriate and sustainable manner consistent with the national fisheries policy and the goals for that particular fishery. Management actions nearly always involve the fisher and hence affect him or her directly. They also influence the abundance, and hence availability, and the size structure of the stocks affected by the fishery. These changes will affect the fisher and other users. Operational objectives for the desired economic and social performance of any management strategy therefore need to be developed, and alternative strategies evaluated against these objectives. The results of the social and economic analyses should be presented to the decision-makers so that they can be considered in making the decision, in the same way as for the biological and ecological information.

In nearly all cases, the quantity of most interest to the fisher is the magnitude of the catch they can expect in the near future, as this is translated directly into income for them. The scientists should therefore attempt to estimate how changes in management measures or strategy are likely to affect the future catches by the fishers. Fishers will probably also be interested in likely changes from year to year in future catches, species and size composition of the catches and, where relevant, in distribution of the fish. These features of the expected future catches can be translated into probable gross income for the fishers, an important item of information for them and for the managers.

Gross income is not the only economic variable that affects fishers' livelihoods, and the costs associated with their fishing activities are as important to them as their income. Different management strategies may affect both fixed and variable costs and hence the total cost and profitability of fishing. Decision-makers need to be aware of the economic and social implications of alternative management strategies and the scientists should include criteria reflecting these consequences and estimate the performance of the different strategies against them. The fishers themselves will be essential sources of this information and should be key participants in the assessment process. However, as with all information, it is important to verify the information obtained from the fishers. In some cases their perceptions may be erroneous, while in others they may see it as being to their advantage to provide incorrect information. Their information should therefore be supplemented from alternative sources as far as possible.

Two examples of economic information that could be useful to the manager in setting general and operational objectives for a fishery are shown in Figure 9. The estimated combined net

---

[2] For further details on Ecosim, see Walters, C.W., V. Christensen and D. Pauly. 1997. Structuring dynamic models of exploited ecosystems from trophic mass-balance assessments. *Rev.Fish Biol.Fish.* 7:139-172 or http://www.ecopath.org/

present value of the shrimp trawling fleets of Trinidad and Tobago and Venezuela, both fishing on the same stock, is shown in Figure 9 a). The results indicate that, from an economic perspective, there is too much capacity in these two fisheries, and that effort should be reduced in both national fleets if an important objective is to maximise net present value. However, there may also be social objectives that need to be considered, for example maintaining employment and earnings per fisher and shore worker. The impact of a reduction in effort on these may also need to be examined and then a decision made which results in a desirable balance between the social and economic goals. In the study from which this figure was taken, the authors provided this information, and some measures of uncertainty, giving the manager very valuable information for identifying, or pursuing the optimal operational objectives.

The implications of having an open access fishery are reflected in Figure 9 b), in terms of the estimated change in rent (gross earnings) per day from the groundfish fishery in Trinidad. The figure demonstrates how, if access is not limited, the gross earnings per unit of effort will decrease with time as more and more fishers enter the fishery, competing for the same fish resources and driving their abundance and yield lower and lower. Such economic and social information is essential to inform the manager on the impacts on the interested parties of different management decisions, including (Figure 9b), a decision to leave things as they are.

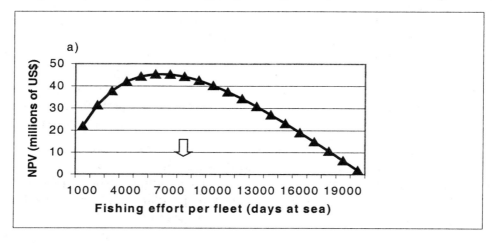

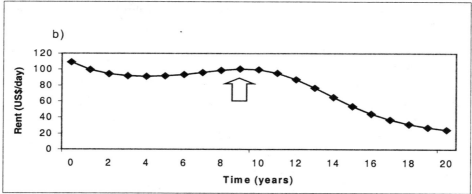

Figure 9. a) Net present value (NPV) of the shrimp landings for the combined trawling fleets of Trinidad and Tobago and Venezuela for different levels of effort. b) The estimated rental obtained per unit of effort in the Trinidad groundfish fishery under open access conditions. In both cases the arrow indicates the current level of effort estimated in those two fisheries. Figures taken from Ferreira, L. and S. Soomai (2000)[3].

---

[3] Management Report: Trinidad and Tobago. In: Report of the 4th Workshop on Assessment and Management of the Shrimp and Groundfish Fisheries on the Brazil-Guianas Shelf. Cumana, September 2000. FAO, Rome (in press).

Different strategies may also have other social implications. For example, in many artisanal fisheries, women and children are involved in processing or selling the landings and changes in management strategies that influence the landings by such fleets may have wider social consequences than the direct impacts on the fishers themselves. Management actions may also have the effect of increasing or decreasing conflicts between different users, and managers should "regulate fishing in such a way as to avoid the risk of conflict among fishers using different vessels, gear and fishing methods" (Code of Conduct, Paragraph 7.6.5). Target and limit reference points should be established for social criteria such as these to enable their consideration in selecting management strategies. As for some other criteria, it may not be always possible to obtain quantitative estimates of the performance of strategies against some of these criteria. and in these cases, the best available qualitative estimate could still provide valuable information. As discussed in Chapter 7, the users themselves form an essential source of such information.

6. THE ROLE OF THE SCIENTIST: PROVIDING OBJECTIVE INFORMATION

The discussion above has highlighted the role of the scientist, which here is taken to include the full spectrum of fisheries scientists including biologist, economist, sociologist, technologist and others, as a provider of scientific information to the decision-makers. If this information is to be useful and to contribute to making the best decisions to achieve the agreed operational objectives, it is essential that the information the scientist provides is accurate, complete and objective. It is then up to the decision-makers to decide on the trade-offs and, where necessary, the sacrifices that need to be made. It is not the task of science to make such policy decisions, science can and should only advise and inform.

Unfortunately, a common problem in fisheries management agencies, and also with other scientists working in resource management, is that the scientists do not always see their role as being limited to the provision of scientific advice, and they may consider themselves to be there to work for a particular cause. In some cases, the management agency may see itself as being there to serve the fishers, perhaps even to serve mainly one section of the fishers, such as the small-scale fishers or the large-scale industry. Under these circumstances, the scientists working for that agency may also adopt, or are pressured to adopt, this partisan role. Conversely, many fisheries scientists are biologists by training and interest and this strong interest in nature can lead them to see themselves as defenders of the resources against a destructive fishing industry.

Any of these prejudices can lead scientists, whether deliberately or through unknowingly succumbing to pressures, to generate advice that is biased towards their interest. For example, they may avoid giving the decision-makers results on strategies that, in their opinion, could result in the resource being driven to too low a level, or they may ignore signs that growth rate or recruitment have been decreasing in recent years to avoid having to recommend a reduction in fishing. Such biases should be avoided at all times. The scientist should not attempt to determine policy or to try to influence policy by manipulation or careful selection of the information he or she presents. Policy decisions should be made transparently and formally by the designated decision-makers in the appropriate forum and, if they are to have confidence that they are making the best decisions, they need to have confidence that they are receiving full and unbaised results from their scientific advisers.

Of course, a scientist may also serve on a decision-making body, where he or she adopts the role of a decision-maker. This is perfectly legitimate, provided the scientist makes it very clear, to themselves and to the others, when they are acting as a scientist, and providing unbiased scientific advice, and when they are expressing an opinion as a member of the decision-making body.

7. HOW SHOULD DECISION-MAKERS AND THE PROVIDERS OF INFORMATION WORK TOGETHER?

When a change to a management strategy is considered, there are obviously many different combinations of management measure, i.e. different management strategies, which could be examined. The selection of which ones to examine in the analyses and which simulations should be undertaken should be a consultative process between the decision-makers and the scientists undertaking the analyses. Only certain changes to the existing strategy may be feasible or desirable, and these should clearly be considered first. There is little point, for example, in considering a management strategy that sets a total allowable catch in a fishery where it is impossible to monitor the catches or landings of all the fishers (see Chapters 4 and 8).

The approach should therefore be to establish first whether any change is necessary (where, for example, an annual TAC is set this may be automatic), and then to discuss which management measures can or should be altered first in an attempt to bring about the desired changes. The scientists can then undertake a series of analyses, using suitable models, to simulate the impact of the new management measures on the resources and fishery. The effects of the changes should be described in terms of the operational objectives and reference points for the fishery.

It may be that the work by the scientists indicates that while the changes in management strategy which have been tested result in some improvements in relation to the operational objectives, none results in fully acceptable results. These results should still be shown to the decision-makers, who may nevertheless compromise and adopt one of the simulation-tested strategies or may request the scientists to undertake further analyses on alternative strategies, seeking one that performs better than those already tested. In this iterative way, the management strategy, or the changes to a strategy, which come closest to achieving the desired results can be identified, making use of the best scientific information available (Figure 10).

8. PRESENTING INFORMATION TO DECISION-MAKERS

Decision-makers in fisheries have to consider several different objectives in deciding on optimal management strategies. Because of potential conflicts in these objectives, there will never be a solution which simultaneously maximises all the potential benefits and minimises all the potential risks. The decisions made will therefore invariably require deciding on suitable trade-offs between these conflicting requirements, and the decisions tend to be of a political nature but, if they are to be good decisions, need to be informed by the best available scientific information.

Enormous advances have been made in stock assessment in recent decades, fuelled especially by easy access to powerful computing capacity. In contrast, formal approaches to decision-making in fisheries have probably progressed very little. There has been an important growth in awareness of the need to involve all key interest groups in the management process, often including in decision-making (see Chapter 7), but once these groups are assembled, the approach to making decisions still usually centres on open debate and argument with all the flaws and problems such an approach involves. The greatest weakness of such informal approaches is that they are heavily influenced by personalities and therefore tend to be subjective and prone to bias arising from, for example, self-interest, short-term objectives prompted by immediate problems, and hidden agendas. Succumbing to any of these may compromise achievement of the agreed long-term strategies and goals.

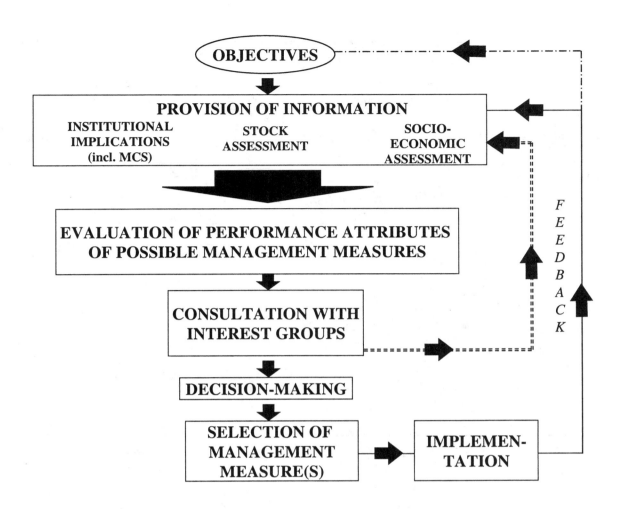

Figure 10. Idealised representation of decision-making in the fisheries management process. Decision-making occurs on a variety of time scales e.g. over days or weeks (double broken-line), annually (solid line) and less frequently e.g. every 3 to 5 years (solid/dashed line).

Some formal statistical methods of decision-making have been developed and some of these have been applied to decision-making in fisheries. Amongst these are multi-attribute analysis, analytic hierarchy process and the use of multi-criteria objective functions. However, these approaches have not proven very popular and do not seem to have been widely applied or adopted for routine application. A likely explanation for their lack of popularity is that they may be perceived to be restrictive and to replace free will with automation. Put another way, they are seen to reduce the opportunity for people to exercise fully their skills in getting their own way!

Where a decision-making body is open to the use of formal statistical methods to assist them in decision-making, these should be used as they have a useful role to play in identifying key issues, setting priorities and minimising the opportunity for hidden prejudices. However, where the decision-makers prefer to operate in the traditional manner of discussion and debate, this has to be recognised and accepted. It seems likely that decisions in fisheries management will continue to be made in (now) smoke-free rooms by committees consisting of individuals representing specific interest groups or selected on the basis of office or expertise. It is essential for these committees to be provided with relevant, objective and easily understood information by the fisheries scientists.

One of the more useful ways in which information can be presented to decision-makers in a form facilitating comparison and decision-making is in a decision table (e.g. Table 5). A well-structured and complete decision-table will not only summarise and present key results from the analyses, but can also serve to remind the decision-makers of their operational objectives, and how different management strategies perform against each of them.

Table 5. Hypothetical decision-table for presentation to decision-makers, enabling comparison of different strategies against some operational objectives. The hypothetical 95% confidence limits are show in parentheses.

Performance Indicator	Management Strategy 1 (existing)	Management Strategy 2	Management Strategy 3
Mean annual biomass of stock as a proportion of unexploited level	36 (18-54)	53 (26-79)	28 (14-42)
Mean annual biomass (proportion of unexploited) of bycatch species most heavily impacted by fishery.	49 (22-66)	63 (28-98)	19 (7-31)
Economic Indicators			
Mean annual catch ('000t)	20 (16-24)	17 (12-22)	23 (17-29)
Mean annual income per fisher ('000 US$).	18 (14-22)	15 (11-19)	20 (15-25)
Inter-annual variability in mean annual income per fisher (% of mean income)	12	9	14
Change in the number of fishers in the fishery compared to the existing level (%)	0	-15	+1

The hypothetical results in Table 5 would present decision-makers with some difficult decisions. They indicate that the present strategy (Management Strategy 1) is having a substantial impact on the target stock, reducing it to an estimated 36% of its mean unexploited level, with a possibility that the stock is as low as 18%. The alternative 2, which could involve a reduction in effort and/or a change in gear selectivity, would have substantial benefits for the target stock and the most heavily affected bycatch stock but would both reduce the average earnings of the fishers and require a reduction of 15% in the number of fishers in the fishery. Strategy 3 would result in a slight increase in the number of fishers and a substantial rise in their average annual earnings in the short-term but with a substantially greater impact on the resource, generating a real possibility of reductions in recruitment (not taken into account in these 'simulations' because of a lack of information) and a downward spiral in biomass and yield. Based on these results, there would be no easy options for the decision-makers in this fishery. Taking the uncertainty into account (in this case that includes the possibility that the stock is as low as 18% of its pristine level), Strategy 2 is clearly the best strategy, and possibly an essential strategy, for ensuring the sustainability of the resource and bycatch species and therefore of the fishery. However, the social and economic impact of Strategy 2 may be considered highly undesirable. Under such circumstances, the decision-makers may choose to go back to the scientists and ask them to attempt to identify alternative strategies that provide a compromise between Strategies 1 and 2, providing adequate protection to the resource but with less severe social and economic implications. This may or may not be possible, but the possibility could be investigated before a final decision was made.

The implications of the different management strategies for the institutional and operational aspects of a fishery should also be considered before final decisions are made. For example, if the decision-makers are considering a choice between managing the fishery purely on the basis of effort regulation or by TAC, the implications for monitoring and control of catches would be important considerations which would have to be brought to the attention of the decision-makers (see Chapter 4). Similarly, the ecosystem effects of a strategy should be considered.

Graphic output, such as that shown in Figures 2 to 9, is usually helpful to the decision-makers. It is essential for the fisheries scientists to communicate with them and find out what sort of information is most useful and how best to present it. Both groups will learn with experience the formats which are most useful. However, this should not be seen as meaning that the scientists should only provide the information requested of them. If they have results or information which they consider important for the decision-makers to see and consider, it is their responsibility to ensure that this information is provided.

Overall, these steps should lead to an approach such as that reflected in Figure 10. An important feature of this figure is the indication of consultation and feedback which should characterise the link between the decision-makers and those providing the information.

9. WHAT ABOUT UNCERTAINTY?

The introduction to this chapter emphasised how big a problem uncertainty, or lack of knowledge, is in fisheries management. Trying to estimate the abundance of fish and their productivity is difficult enough, and the estimates we obtain of these values are always just that, estimates, with considerable uncertainty associated with them. When we try to forecast or predict what the abundance of fish will be next year we introduce even more uncertainty, and when we try to forecast how the stock, fish community or fishers will respond to management actions we introduce even more.

There are many sources of uncertainty in fisheries stock assessment and management and these can be summarised as[4]:

- *process uncertainty*, or random variability, in the biological and ecological processes themselves, such as in recruitment to a stock;

- *observation uncertainty*, from attempts to measure factors such as total catch, biomass (e.g. through a survey), or effective effort in a fishery;

- *estimation uncertainty* in our final estimates of quantities, such as the status of the stock or B_{MEY}, arises from process and observation uncertainty above and also because our models are usually simplifications of the true ecological processes;

- *implementation uncertainty* arising in the implementation of management measures, including how effective they will be and how well the fishers will comply with them; and

- *institutional uncertainty* which refers to the uncertainty in how well participants in the process can communicate with each other, to what extent people are willing to compromise and how the scientific information is understood, all influencing how decisions will be made and therefore how good those decisions will be.

We can estimate values for some of these uncertainties and use these values in stock and bioeconomic assessments and decision-making. For example, by measuring recruitment to a stock over a number of years, an estimate can be obtained not only of the average recruitment but also of the variability from year to year, which could be expressed by measures such as the

---

[4] From Francis, R.I.C.C. and R. Shotton. 1997. "Risk" in fisheries management: a review. *Can. J. Fish. Aquat. Sci.* **54**: 1699-1715.

standard deviation about the mean, the 95% confidence intervals of the mean or simply the range of observed recruitments. Similarly, it may be possible for the uncertainty in abundance, biomass or the potential yield from a resource to be estimated. For example, the best estimate of the biomass of a stock of sardine, based on a hydroacoustic survey or a biomass dynamic model, may be 100 000 t but when the uncertainties are calculated, it is found that the 95% confidence limits of the estimated biomass are from 60 000 t to 140 000 t. This means that the <u>most likely</u> estimate is 100 000 t but that there is 95% certainty that it lies between 60 000 and 140 000 t. Ninety-five percent confidence limits of at least 40% on either side of the mean, as used in this hypothetical example, would be typical of many estimates from well monitored fisheries.

In some cases, good numerical estimates of uncertainty may not be available, but the scientists should then provide a carefully considered statement of how good their estimate is. For example, they could indicate whether their estimate of total landings is very good, good, reasonable or only an approximation. Implementation uncertainty and institutional uncertainty are generally much more difficult to estimate than the other types of uncertainty listed above. In most cases the best information that may be available for these types is, for example, that there is a high or a low probability of serious violation of regulations (see Section 5 of Chapter 8), or that there is a high, medium or low level of confidence that the management measures selected will achieve the desired result. Even information such as this will assist the decision-makers in interpreting the information and making the best decisions.

In the past, fisheries management tended to ignore the uncertainties and act on the best estimates as being the correct answers. However, with increased computing power and greater understanding of how much we don't know in fisheries management, the modern tendency is to try to estimate the various uncertainties (risk assessment) and to consider them in determining and implementing management measures and strategies (risk management). Risk assessment is discussed in more detail in an Appendix to this Chapter.

Risk management is still in its infancy in fisheries management and there are no commonly applied formal ways of doing it. In essence, risk management requires the decision-makers to make the best decision they can based on the information they have but then also to consider the likelihood of that decision being wrong. They should then consider modifying the decision, such as the selected management measures, so that the strategy will not only work well if circumstances and behaviour fall within the expected range, but also so that it won't go too badly wrong if circumstances and behaviour turn out to be very different from the initial expectations. More formally, this can be referred to as making decisions which are robust to the uncertainties. Robustness testing requires the use of models and information to consider how a management strategy will perform under different conditions or states of nature to those considered in the basic assessments, or how it will perform if some of the information used in those basic assessments is incorrect. It provides a means of identifying possible undesirable outcomes from a management strategy before they occur, thereby allowing modifications to be made to the strategy before it is implemented to try to avoid such outcomes. In other words, robustness testing is a means of reducing, in advance, the probability that the selected management strategy will go badly wrong, or ensuring that it can quickly be adapted, if the ecosystem or the fishery or both do not behave in the way they were expected to when the strategy was designed. Being robust to uncertainty could mean being more cautious than the basic assessments suggested in, for example, setting total allowable effort. Alternatively, it could involve ensuring that effort can rapidly be scaled down, without creating unnecessary social and economic disarray, if production by the resource was less than had been expected when the assessments were undertaken.

This approach has to be balanced, of course. If one takes an extreme view of uncertainty and robustness, then the only way to minimise risk in the face of the inevitable uncertainty is to close the fishery. This is not a practical approach and the management strategy should be designed to be robust only to changes which could reasonably be expected to occur.

Including consideration of uncertainty in assessment and management does put much greater demands on all involved in the management process, including scientists, fishers and other interested parties, managers and decision-makers. However, it also results in a much greater chance of good decisions being made and management strategies being implemented that stand the greatest chance of achieving the objectives. Few individuals would set out on a long journey without taking a spare tyre, maps and some extra money in case of unexpected but possible problems. Fisheries management requires, at a minimum, the same level of caution, or risk management.

10. UNCERTAINTY AND THE PRECAUTIONARY APPROACH

Some guidance has been provided in the earlier sections of this chapter on how uncertainty can be taken into account in making decisions but, except in the case of some formal and rigorous statistical approaches, there are no widely accepted and applied methods for incorporating knowledge of uncertainty into decision-making. There has, however, been a lot of discussion about this and this discussion has been reflected in a general philosophy or concept known as the precautionary approach. The application of the precautionary approach in fisheries is included in the Code of Conduct, and is the subject of one of the FAO technical guidelines to the Code of Conduct (FAO, 1996).

The precautionary approach in fisheries management can best be summarised as "the application of prudent foresight" (FAO, 1996). FAO (1996) goes on to list the requirements of the precautionary approach as including (Paragraph 6):

- "consideration of the needs of future generations and avoidance of changes that are not potentially reversible;
- prior identification of undesirable outcomes and of measures that will avoid them or correct them promptly;
- that any necessary corrective measures are initiated without delay and that they should achieve their purpose promptly.....;
- that where the likely impact of resource use is uncertain, priority should be given to conserving the productive capacity of the resource;
- that harvesting and processing capacity should be commensurate with estimated sustainable levels of resource (production), and that increases in capacity should be further contained when resource productivity is highly uncertain..."

It is also suggested (Para. 7d) that, in applying the precautionary approach, "the standard of proof to be used in decisions regarding authorization of fishing activities should be commensurate with the potential risk to the resource, while also taking into account the expected benefits of the activities."

These are important considerations and the reader is urged to study carefully the FAO Technical Guideline on the Precautionary Approach and the relevant sections in the Code of Conduct (Sub-article 7.5).

11. CONCLUSIONS

A consistent theme running through this Guidebook is that fisheries management is a complicated task with broad goals that are usually in conflict but which need to be reconciled through the formulation of operational objectives which aim to provide benefits for society in a sustainable way. The fisheries manager is responsible for seeing that this is done and for ensuring that management strategies are developed and implemented which stand the best chance of achieving these reconciled objectives. There are many tools for doing this but, because of the complexity of ecosystems and their interactions with fisheries, the information required for making the decisions is usually incomplete and includes a lot of uncertainty. This chapter has attempted to describe what sort of information the manager should be asking for from the scientific branch of the management agency, how that information should be presented to the decision-makers and how they should use it in making their decisions.

The most important aspects of this process are that only well-informed decision-makers can make good decisions. Therefore, the best information available, given the staff and resources available to the agency, should be used to advise the decision-makers. It is the responsibility of the scientists to ensure that they are collecting appropriate information to provide the necessary advice, that they store this information in a way which makes it easily accessible in the future, they analyse it using appropriate methods, and provide easily understood, complete (as far as is possible) and unbiased information that is relevant to the decisions that have to be made.

The examples of methods and approaches presented in this chapter are just some of the types of questions which can be expected to arise in fisheries management and of the types of scientific information that may help to answer them. Further information can be found in the fishery assessment books already referred to, as well as in the vast numbers of scientific papers on fisheries assessment and management which are published every year. It is very important that all fisheries management authorities have access to staff familiar with at least the standard approaches to the types of analyses presented here. Without this, properly informed decisions and therefore effective and responsible use of the fishery resources will not be possible.

Good communication is important at all levels, and the decision-makers, scientists and other interested parties should be working together to ensure that the correct information is being provided, and that it is being interpreted properly. Following all these steps will not guarantee that the correct decisions are made, but it will help to ensure that the best decisions are made given the information and resources available. That is all that can be asked of the decision-makers and those whose task it is to provide them with the information they require.

12. RECOMMENDED READING

Caddy, J.F. and R. Mahon. 1995. *Reference points for fisheries management.* FAO Fisheries Technical Paper **382**. 83pp.

Charles, A.T. 2001. *Sustainable Fishery Systems.* Blackwell Science, Oxford, United Kingdom.

FAO. 1996. *Technical Guidelines for Responsible Fisheries* No. 2: Precautionary Approach to Capture Fisheries and Species Introductions. FAO, Rome. 54 pp.

FAO. 1997. *Technical Guidelines for Responsible Fisheries* No. 4: Fisheries Management. FAO, Rome. 68 pp.

FAO. 1999a. *Guidelines for the routine collection of capture fishery data.* FAO Fisheries Technical Paper **382**. 113pp.

FAO. 1999b. *Technical Guidelines for Responsible Fisheries* No. 8: Indicators for Sustainable Development of Marine Capture Fisheries. FAO, Rome. 68 pp.

Gayanilo, F.C., Jr. and D. Pauly (eds). 1997. *FiSAT. FAO-ICLARM stock assessment tools. Reference manual.* FAO Computerized Information Series. FAO, Rome.

Hilborn, R. and C.J. Walters. 1992. *Quantitative Fisheries Stock Assessment. Choice, Dynamics and Uncertainty.* Chapman and Hall, New York. 570pp.

Seijo, J.C., O. Defeo and S. Salas. 1998. *Fisheries bioeconomics. Theory, modelling and management.* FAO Fisheries Technical Paper **368**. 108pp.

Sparre, P. and S.C. Venema. 1992. *Introduction to tropical fish stock assessment. Part 1 - Manual.* FAO Fisheries Technical Paper **306/1**. 376pp.

Sparre, P.J. and Willmann, R. 1993. *BEAM 4. Analytical bioeconomic simulation of space-structured multi-species and multifleet fisheries*. FAO Computerized Information Series. FAO, Rome.

Punt, A.E. and R. Hilborn. 1996. *BIODYN. Biomass dynamic models. User's manual*. FAO Computerized Information Series. FAO, Rome.

APPENDIX: RISK ASSESSMENT

What is a risk assessment?

Risk assessment is usually undertaken by the scientific staff of a fisheries management agency and should include not only assessment of biological risks but of economic and social risks as well. As with all fisheries assessment, risk assessment should be directly related to the operational objectives.

Defining Risk

Risk is commonly defined as the probability of something undesirable happening, but in making use of risk it is necessary to be more precise and the undesirable events must be decided on and quantified. Those of particular concern will relate to the fishery operational objectives e.g. the stock falling below a minimum threshold level, the income to the fishery as a whole or by sub-sector falling below a certain level, or total number of employee days or jobs being reduced below a specified threshold.

In risk assessment, consideration also needs to be given to what is considered an acceptable level of risk for each performance indicator by the different interested parties. There are no rigid guidelines for deciding on an acceptable level of risk for a stock or stocks, and this must represent one of the greatest areas of potential disagreement and hence of weakness, in resource risk assessment (Butterworth et al. 1997). However, an appropriate threshold level of risk should be able to be identified by, for example, comparison with the level of risk for an event in an unexploited population or during some previously observed period when it was considered productive. When considering sustainability of a resource, a fundamental measure of risk should be the probability of recruitment failure brought about by low spawner biomass and, clearly, this risk should not be allowed to be too high.

Meaningful economic and social measures of risk, such as those related to income and employment, will also be difficult to agree on and to define. While the choice of a threshold to avoid crossing may be relatively easy, such as avoiding making a loss or avoiding any reduction in employment, it may be more difficult to agree on the point at which the risk of crossing these thresholds becomes too high. However, in contrast to defining biological risks, the question is amenable to debate with those most directly affected, the fishers and other interested parties, who should be directly involved in selection of the acceptable risk levels. In addition, it should be easier to determine the consequences of crossing any social or economic threshold than it is for those of falling below some stock biomass threshold.

An integral part of determining an acceptable (or unacceptable) risk or probability of crossing a threshold is the time period over which the risk is measured. For example, the risk of being struck by lightning is partly dependent on the length of time the target is exposed to lightning. Everything else being equal, the risk of being struck by lightning within a ten-year period is ten times the risk of being struck in a one-year period. Discussion on what is an acceptable level of risk must therefore include clear definition of the length of time over which the risk is measured. Where risk is being measured using models, as it normally will be, this means considering the time period over which the model is projected. This should take into account the dominant time-scales of the stock, particularly average life span of the resource, and the fishery (e.g. life span of a vessel etc). Periods of between 10 and 20 years are frequently used in estimating risk in fisheries.

Estimating Risk

Risk assessment is usually undertaken using the standard stock assessment approaches in conjunction with the available data on the resource or resources and the fishery. The first step is to estimate the important parameters and variables which describe the dynamics of the resource and fishery, and the uncertainty or error in these estimates, including the distribution (e.g.

whether a normal, log-normal or uniform distribution) and magnitude of the distribution of errors. These estimates can then be used to construct a forecasting model of the fishery-resources system. The type of model to be used will depend on the questions to be asked and the parameters and variables which have been estimated. For example, at a more simple level, it may only be possible to use per-recruit models to investigate the impact of different levels of fishing mortality and different ages at first capture on relative yield and biomass-per-recruit. Alternatively, if there are estimates of biomass and recruitment as well, it may be possible to estimate average yield and inter-annual variability in yield under different management strategies. If there are only data on catch and effort, a biomass dynamic approach could be used. The same class of model as was used to estimate parameters and their errors should normally be used for forecasting the impacts of management strategies. The model is then used in forecast mode to investigate the impact of, for example, different catches, levels of effort or gear type on biomass, size or age structure and average yield given the estimated uncertainties. The models may, and normally should, also include estimation of social and economic performance, including uncertainty, to allow each possible management strategy to be evaluated on its performance across all operational objectives.

The uncertainty estimates are used in the models which are run in a stochastic or probabilistic mode i.e. the model is run many times in a Monte Carlo manner for each management strategy being tested. Typically the model is run between 1 000 and 10 000 times for each strategy, drawing the values of selected parameters for each model run from a probability distribution defined by the error distribution of the parameter estimates (instead of keeping them all constant). Therefore different parameter values are used in the model in each run, generating different results. When the runs are completed, the average values of the performance indicators, reflecting the operational objectives, and their range or distribution can be calculated. For a risk assessment, the number of runs in which the performance indicator of interest fell outside the selected risk threshold (i.e. the number of times each undesirable event happened) can then be counted, giving an estimate of the risk under the management strategy being simulated.

This sort of analysis leads to information of the type shown in the Table below in the form of a decision table. This example was taken from simulations used to assist in the selection of a management strategy (more formally a "management procedure" see Cochrane et al. 1998 for definition of a management procedure) for the South African anchovy fishery. The table shows the performance indicators considered important in this fishery: biological risk to the resource (kept constant at 30% in this example), average annual catch and catch variability. The results shown here allowed the decision-makers (the managers and the fishery interested parties) to weigh-up the trade-offs between maximising average annual catch and stability (including the minimum TAC) which would be best for the efficient management of the fishery. It had already been agreed that 30% risk of biological "failure" was acceptable (using a particular definition of risk as described in the table) but if this was controversial, the simulations could be repeated for different levels of risk to show the trade-offs between changing biological risk and the average catch and catch variability.

Table. Example of performance measures for different management options for the South African anchovy. For all options, the risk of the stock falling below 20% of unexploited biomass within a 20 year period equals 30%. "Management Options" shows aspects of the rules used in setting the total allowable catch (TAC) each year. "Max. reduction" = the maximum reduction in TAC from one year to the next. From Butterworth *et al.* 1992.

Management Option	Performance indicators	
	Annual Average Catch ('000 t)	Interannual Catch Variability (%)
'Base Case (BC)' Max. TAC = 600 000 t Min. TAC = 200 000 t Max. reduction between years = 40%	315	25
BC but max. TAC = 450 000 t	314	23
BC but min. TAC = 150 000 t	328	25
BC but max. reduction = i) 50% ii) 25%	 321 285	 25 22

REFERENCES

Butterworth, D. S., De Oliveira, J. A. A., and Cochrane, K. L.. 1992. Current initiatives in refining the management procedure for the South African anchovy resource. *In Proceedings of the international symposium on management strategies for exploited fish populations* Ed. by G. Kruse, D. M. Eggers, R. J. Marasco, C. Pautzke, and T. J. Quinn. Alaska Sea Grant College Program AK-SG-93-02. 439-473.

Butterworth, D. S., Cochrane, K. L. and De Oliveira, J.A.A.. 1997a Management procedures: a better way to manage fisheries? The South African experience. In Pikitch E.K., Huppert D.D. and Sissenwine M.P. eds. *Global Trends : Fisheries Management* (Proceedings of the Symposium held at Seattle, Washington, 14-16 June, 1994). Bethseda, Maryland : American Fisheries Society Symposium 20, 83-90.

Cochrane, K. L., Butterworth, D. S., De Oliveira, J. A. A., and Roel, B. A. 1998. Management procedures in a fishery based on highly variable stocks and with conflicting objectives: experiences in the South African pelagic fishery. *Reviews in Fish Biology and Fisheries,* **8**: 177-214.

CHAPTER 6

USE RIGHTS AND RESPONSIBLE FISHERIES: LIMITING ACCESS AND HARVESTING THROUGH RIGHTS-BASED MANAGEMENT

by

Anthony T. CHARLES[1]

Saint Mary's University, Nova Scotia, Canada

1. WHAT ARE USE RIGHTS? 131
2. WHY ARE USE RIGHTS RELEVANT TO FISHERY MANAGEMENT? 133
3. HOW DO USE RIGHTS RELATE TO OTHER RIGHTS IN FISHERIES? 134
4. WHAT FORMS OF USE RIGHTS ARE THERE? 137
   - 4.1 Territorial Use Rights 138
   - 4.2 Limited Entry 139
   - 4.3 Effort Rights (Quantitative Input Rights) 140
   - 4.4 Harvest Quotas (Quantitative Output Rights) 141
5. HOW ARE USE RIGHTS IMPLEMENTED? 143
   - 5.1 Are use rights already in place? 143
   - 5.2 What is the 'best' set of use rights? 143
   - 5.3 What is the underlying policy framework? 144
6. SYNTHESIS 151
7. REFERENCES AND RECOMMENDED READING 155

1. WHAT ARE USE RIGHTS?

Elsewhere in this volume, various forms of fishery regulation are discussed – such as area closures, limited entry and other input controls (effort limitation) and output controls (quotas). These regulations address a range of fishery issues: Who can go fishing? Where is fishing allowed? How much gear can be used? How much fish can be caught? Suppose, however, that we look at these restrictions from a different perspective, namely that of *use rights* – the *rights* held by fishers or fishing communities to *use* the fishery resources.

Whenever a fishery is managed by restricting who can have access to the fishery, how much fishing activity (fishing effort) individual participants are allowed, or how much catch each can

---

[1] This chapter draws on earlier work appearing in Townsend and Charles (1997) and Charles (2001). I am grateful to Ralph Townsend, Melanie Wiber and Parzival Copes for many helpful discussions, but any errors herein remain my responsibility.

take, those with such entitlements are said to hold use rights. Such use rights are simply 'the rights to use', as recognised or assigned by the relevant management authority (whether formal or informal). For example, limited entry – seen as an 'input control' from the perspective of resource management – can be viewed as an 'access right' from the perspective of fisher management. Certain individuals, groups or communities have the right to 'use' the fishery (i.e., to go fishing) while all others do not have that right. Similarly, limits on the number of traps that are allowed to be used might be seen as a (negative) restriction, or as a (positive) use right – the fisher, group or community has the right to use a certain number of traps.

Naturally, along with rights go responsibilities: as the FAO Code of Conduct for Responsible Fishing (Paragraph 6.1) notes, "The right to fish carries with it the obligation to do so in a responsible manner...". A key aspect in moving toward responsible fisheries thus lies in developing effective and accepted sets of both rights and responsibilities among fishers. To this end, the present chapter focuses on use rights, exploring the various forms of such rights, their advantages and disadvantages, policy issues relating to the choice among alternative use rights systems, and issues concerning how use rights are implemented in practice, and who can or should hold these rights.

Use rights options range widely; for example, each of the following approaches to fishery management involves use rights:

- Customary Marine Tenure (CMT) and Territorial Use Rights in Fishing (TURFs) have long been applied by indigenous communities in determining for each member of the community (whether a fisher or household) the location where that member can access fishery resources;
- limited entry was the initial approach to use rights in modern 'state' management of fisheries, providing a limited number of individual fishers with the right to access the fishery;
- quota allocations made to individual fishers, companies, cooperatives, communities, etc., to catch a specified amount of fish, are numerical (quantitative) use rights, as are allocations of rights to a certain level of fishing effort (e.g., quantity of gear or days fishing).

While there is considerable diversity in use rights systems, they can generally be placed within two major categories:

(a) access rights, which authorise entry into the fishery or into a specific fishing ground;

(b) withdrawal (harvest) rights, which typically involve the right to a specific amount of fishing effort (e.g., to fish for a certain amount of time or with a certain amount of gear) or the right to take a specific catch.

Each of these categories can occur at various organisational levels, i.e. rights held by individuals, by communities or regions, or by specific groupings such as fishing vessel or gear sectors. Indeed while use rights are often discussed in terms of individual fisher rights, an important form of use right, both historically and currently, is that held collectively by a community.

Note that use rights arise in a multitude of contexts well beyond the fishery. For example, consider the owner of a home in a setting such as a rural village or an urban condominium (an apartment block in which each unit is privately owned). Such homeowners certainly have the right to 'use' the home. In addition, as will be discussed below, they likely have other rights as well: the right to exclude others from using the home, and perhaps the right to sell the home to someone else. Now consider the common areas surrounding the home, and neighbouring homes, such as the village pasture or the garden outside the condominium building. The group of homeowners may well share the use rights over these common areas, with no single person having the right to exclude others nor to sell the common area. Similar situations arise in a wide variety of settings in which individuals and families hold various rights within their own households, as well as shared collective rights over community property. Alternatively, in such a situation, the cultural context may be such that collective (group) rights predominate, as in some indigenous/native societies.

2. WHY ARE USE RIGHTS RELEVANT TO FISHERY MANAGEMENT?

The Code of Conduct (Paragraph 10.1.3) makes reference to use rights, not only within fisheries but pertaining to coastal resources in general: "States should develop, as appropriate, institutional and legal frameworks in order to determine the possible uses of coastal resources *and to govern access to them* taking into account the rights of coastal fishing communities...". Why are use rights so important?

Use rights aid management by specifying and clarifying who the stakeholders are in a certain fishery, while also aiding these stakeholders – whether fishers, fishers' organisations, fishing companies or fishing communities – by providing some security over access to fishing areas, use of an allowable set of inputs, or harvest of a quantity of fish. If use rights are well established, fishers know who can or cannot access the fishery resources, how much fishing each is allowed to do, and how long these rights are applicable.

Fisheries with clearly-defined use rights may be contrasted with open access fisheries. In their fullest form, open access fisheries are ones in which there is no regulation of the fleet or the catch. In particular, there are no limits to access – anyone can go fishing. Perhaps the most famous cases of open access (and, until recently, its most serious manifestation) have been the high seas fisheries – taking place in ocean spaces located outside any single nation's jurisdiction.

It has become accepted wisdom, based both in theory and in the experiences of fishery collapses and stock depletion world-wide, that such open access will lead to likely-disastrous conservation and economic problems. Unregulated 'laissez-faire' (free enterprise) exploitation of marine resources is among the greatest threats to the long-term sustainability of fisheries. Indeed, the threat posed by such open access fisheries was a major factor leading to efforts to regulate fisheries on the high seas, through the United Nations Conference on Straddling Fish Stocks and Highly Migratory Fish Stocks and the subsequent United Nations Fish Stocks Agreement[2] implementing the relevant provisions of the UN Law of the Sea.

Note that the term 'open access' is also used sometimes to refer to a fishery in which there are no controls over the number of boats or the amount of gear, even though the total catch may be regulated. In such a case, the fish stocks may not necessarily collapse (if regulations work) but the fleet may become excessive (over-capitalised), driven by economic incentives to enter the fishery and to invest in larger boats (in a 'race for the fish' in which those who catch the most fish first are the 'winners'). With more inputs used than necessary to catch the fish, the economic health of the fishery may be threatened even if the resource is safeguarded.

With open access fisheries having a bad name both internationally and within national jurisdictions, the overall need for and desirability of restricting access is usually accepted as a basic premise in fishery management. Indeed, the need for use rights – specifically access restrictions – has long been understood in many parts of the word. Informal and traditional use rights have existed for centuries in a wide variety of fishery jurisdictions. Even in cases where direct government regulation of fisheries dominates, use rights are being implemented with increasing frequency.

---

[2] Agreement for the Implementation of the Provisions of the United Nations Convention on the Law of the Sea of 10 December 1982 relating to the Conservation and Management of Straddling Fish Stocks and Highly Migratory Fish Stocks.

Use rights are relevant to the fishery manager not only in resolving open access problems, but also in helping to clarify who is being affected by management. This has the following benefits.

- First, an effective use rights system eliminates (or reduces) the need for fishery management as such to deal with one major element of complexity and uncertainty – that of identifying the set of users and regulating that group. In an already complex and uncertain management system, this can be a major benefit.

- Second, when use rights are clear, fishers and fishing communities can better plan their resource harvesting, with users better able to maximise the value of the output within a conservationist framework, and to adapt to changing conditions. Furthermore, use rights may assist in reducing the magnitude of conflict in fisheries (in keeping with Paragraphs 7.6.5 and 10.1.4 of the Code of Conduct). These factors are helpful in enhancing the fishery's overall resilience – its ability to 'bounce back' from unexpected shocks.

- Third, fishery management can more easily identify use rights holders as those needing to meet certain conservation requirements. For example, the Code of Conduct (Paragraph 6.6) states that "where proper selective and environmentally safe fishing gear and practices exist, they should be recognized and accorded a priority…". This implies the adjustment of use rights to promote (or favour) certain gear types or fishing practices.

- Fourth, with clear-cut use rights, conservation measures to protect 'the future' become more compatible with the fishers' own long-term interests, which may encourage adoption of a conservation ethic and responsible fishing practices, and greater compliance with regulations (Code of Conduct, Paragraph 6.10). As noted earlier, the Code of Conduct (Paragraph 6.1) highlights the necessity of such connections between use rights and conservationist practices. (This also points to the need for care in establishing use rights, since it is possible that certain use rights could be accompanied by anti-conservationist incentives, leading to such actions as high-grading, the discarding of low-valued fish in order to maximise profits.)

3. HOW DO USE RIGHTS RELATE TO OTHER RIGHTS IN FISHERIES?

Use rights are put in place to specify who is to be involved in resource use, thereby making management more effective and conservation more likely. Is there also a need to specify who is to be involved in fishery management? In past times, this question may have seemed quite irrelevant – management was done by the managers, typically government officials within a commercial fishery context. Over time, however, it has become clear that fishery management is rarely successful when practised in a top-down manner, because the manager rarely if ever has the time and finances to fully monitor thousands of fishers at sea. Thus effective management requires the support (or at least acceptance) of fishers, accompanied by some degree of self-regulation.

This has led to the emergence of new co-management arrangements involving joint development of management measures by fishers, government and possibly local communities. Chapter 7 focuses on this topic, which has been the subject of considerable study in recent years. In the language of fishery rights, co-management requires allocation of management rights, the right to be involved in managing the fishery. Note that management rights and use rights can be seen as parallel forms: the former specify the right to participate in fishery management just as the latter specify the right to participate in the fishery itself. Management rights reflect the need, as noted in the Code of Conduct (Paragraph 6.13), to "facilitate consultation and the effective participation of industry, fishworkers, environmental and other interested organizations in decision-making with respect to the development of laws and policies related to fisheries management…".

Who should hold management rights? The above discussion suggests that, if only for pragmatic reasons, fishers (those with use rights) should be among the rights-holders. The government – with responsibility, as is usually the case, to conserve the resource, to produce benefits from that resource, and to suitably distribute those benefits – will hold management rights as well. To what extent should management rights also be held by communities, nongovernmental organisations (NGOs) and the general public? This is an important question, the answer to which may vary depending on the level of management under discussion, and which is discussed further in Chapter 7.

Consider first the operational or tactical level of management – involving measures such as closed areas, closed seasons, and allowable hook or mesh sizes, that affect the fishing process directly. At this level, it is particularly crucial for fishers to hold management rights, so as to encourage compliance at sea. However, there may often be less interest among communities, NGOs and the general public in these detailed operational aspects. (An exception may be cases in which ecosystem protection is an issue.) On the other hand, debates over strategic management – concerning the fishery's overall objectives and policy directions – are typically matters of public interest, in which the general public, and fishing communities in particular, are legitimate interested parties. Thus, a wide spectrum of interested parties will (should) hold management rights in dealing with strategic management issues, and in setting objectives for use of the fish resources and of the ecosystem as whole. This is increasingly the case with small-scale community-based fisheries; for example, recent legislation in the Philippines places management rights over coastal 'municipal fisheries' clearly at the level of the local municipality (Congress of the Philippines, 1998).

Management rights are one of three types of 'collective choice' rights (as identified by Ostrom and Schlager 1996), the other two being exclusion rights (the right to allocate use rights, and thereby determine who can access the fishery) and alienation rights (the right to authorise the transfer or sale of other rights). These collective choice rights may be held by both users and non-users, contrasting with use rights which essentially are held only by fishery users. For example, while a fishing community may not hold use rights *per se*, it may have management rights (as discussed above) as well as exclusion and alienation rights (relating to decisions about allocating and/or selling off use rights). Precisely who should hold management, exclusion and alienation rights, and what institutions are suitable to deal with such rights, are becoming major issues, likely to receive increasing attention in the years ahead. (See Chapter 7, for further discussion of this matter.)

The more of the various types of rights are held, the more complete is the set of rights. For example, a fisher who owns a fishing boat likely has a right to use the boat, as well as an exclusion right (to prevent others from using the boat), and an alienation right (to sell the boat). On the other hand, a fishing license provides a use right to the fish resources, but likely not an exclusion right to prevent others from using the fish stocks. Thus fishers typically have more complete rights over their boats than over their use of the resource.

Furthermore, it is crucial to recognise that a fisher holding use rights has the right to access the fishery, but the fisher does not own the fish *per se* until those fish are actually caught. Thus use rights do not imply ownership of the resource itself. Unfortunately, this crucial distinction has been confused at times, with use rights (such as individual quotas) promoted by suggesting that fishers holding these rights will in fact 'own' fish in the sea, just as one may own their fishing boat. This idea has been at the root of much recent conflict in fisheries, often between users and non-users, but is not at all what is meant by use rights.

Indeed, in this regard, it is useful to compare the fishery with other natural resource sectors, where the difference between resource access (in the form of use rights) and resource ownership is perhaps clearer. Consider the case of forestry. In jurisdictions with significant government-owned forest, it can be standard practice for industrial harvesting companies to hold leases on specified areas of forest. The companies do not own these forests, but they do have the right to use the resources, often subject to conditions, e.g. that reforestation accompanies harvesting, so as to ensure sustainability. Similarly, in the oil and gas sector, the focus is more on the use right to a particular oil field. The use right *per se* may be 'owned' but ownership of the resources on or in the ground is not in question.

What are 'Property Rights'?

Use rights, management rights and the other types of rights described in this section all fall under the broad heading of property rights. Property rights describe relationships between people over various forms of property. Consider, for example, two such units of property: a fishing boat and a fish stock. With fishing boats, we typically see property rights as being reasonably clear: the 'owner' of the fishing boat can use the boat (use rights), prevent others from using it (exclusion rights), and sell it if desired (alienation rights). Other people have no 'property rights' over the boat. On the other hand, property rights over fish in the sea are typically less clear – as described earlier, different sets of people may hold use rights, management rights and exclusion/alienation rights. This reflects a common focus of property rights analysis on comparing 'bundles' of rights associated with units of property.

Many publications have appeared on property rights in fisheries, from a variety of perspectives. Fishery managers therefore are likely to come across discussions of property rights at some point. However, it should be noted that in most cases, fishery management will focus on issues of access, harvesting and management itself – which involve use rights and management rights specifically, rather than the more nebulous topic of property rights. It seems as well that use of the term 'property' tends to create conflict, with allocations of use rights over fish in the sea misinterpreted as implying 'ownership' of the fish. This conflict arises because, even if fishers hold use rights, the fish in the sea do not belong to those fishers until they are caught. Who owns the fish while they are swimming freely? Property rights theory does provide some help here, by describing four possible 'property regimes' that could apply to fish in the sea.

Non-Property. Traditionally, high seas fish stocks were no one's property. The fish were there for the taking. No one could claim ownership and exclude others. This represented a lack of property rights, a case of 'non-property'. As time has passed, fewer and fewer of the world's fishery resources have been exploited in the absence of property rights.

Private Property. As noted earlier, whenever a fisher catches a fish, once it is brought out of the water into the vessel, that fish became the private property of the fisher. Even when fish are still swimming in the water, they may be private property. In some nations, the fish in a river that passes through private land can be the private property of the land owner. The same could be said for fish in a lake located entirely on someone's private land. In such cases, only the owner of the resource has the right to decide the use of the resource – subject possibly to societal constraints, such as those that may be imposed to preserve biodiversity.

State Property. In many countries, fish in the oceans within the State's EEZ are the property of the nation's citizens, and managed on their behalf by the government. In such cases, the fish are said to be 'state property'. The fish become private property when caught, but remain state property as long as they are in the sea. Typically such resources cannot be privatised without legislation, or perhaps even constitutional change.

Common Property. Suppose that the fish in the sea are 'owned in common' by a certain identifiable group of people – e.g., the set of citizens within a specific local jurisdiction, such as a coastal community, or the members of a native tribe, but not a single private individual or a company. Suppose further that the fishing activity of any one fisher detracts from the welfare of others, and it is difficult for the fishers, as a group, to exclude other potential users. In such circumstances, the fish are referred to as 'common property' – a regime that is pervasive world-wide (although, until recently, little studied relative to the other property regimes above). Note that the common property concept arises in two different modes. In most studies, the relevant group of people holding the common property is relatively small and well-defined (such as a community). On the other hand, in common usage, the fishery resources of an entire nation are often referred to as common property – in which case the 'group' is defined as being so large as to include all citizens of a nation, and common property is equated to state property, as defined above.

Note that use rights can be implemented under any of private property, state property and common property (and to some extent even in the case of non-property, e.g. the United Nations Fish Stocks Agreement (Article 10), provides the capability to prescribe use rights on the high seas). Conversely, a lack of use rights – i.e., open access – lies at the root of most problems with resource depletion, whatever the property rights regime.

The latter point helps to clarify the common confusion between 'common property' (property held in common by a group) and 'open access' (a lack of use rights to limit access to the property). Indeed, this confusion has led to a mistaken belief that common property fisheries are necessarily open access, and therefore destined for over-exploitation – a confusion arising out of a famous 1968 'Tragedy of the Commons' article by Hardin. In reality, a great number of studies have shown that while some 'common property' fisheries are indeed open access, this is by no means the rule. The real issue in a common property fishery is whether the rights holders collectively can develop an effective management institution, something that has in fact occurred in a great number of cases world-wide.

4. WHAT FORMS OF USE RIGHTS ARE THERE?

This section examines in turn the various forms of use rights, fitting within the headings given earlier:

- *access rights*, which permit the holder to take part in a fishery (limited entry) or to fish in a particular location (territorial use rights or 'TURFs');
- *withdrawal rights*, which typically involve quantitative (numerical) limits on resource usage, either through input (effort) rights or output (harvest) rights.

These various forms are depicted diagrammatically in Figure 1.

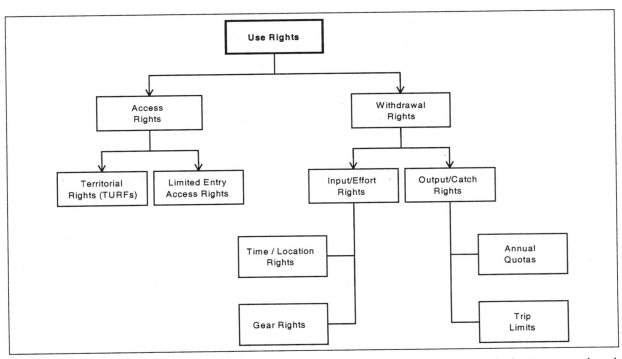

Figure 1. The relationships between the different forms of use right. (See the article by Townsend and Charles (1997) for further details.)

Two points are in order here. First, most of the use rights options to be described here correspond to an input or output control discussed in Chapter 4; they are really the same thing, seen from differing perspectives. For example, limited entry – seen as an 'input control' from the perspective of resource management – corresponds to an access right from the perspective of fishers management. Similarly, the fisher might see a control placed on the number of traps he or she is allowed to use in a fishery as a (negative) restriction, or as a (positive) use right. Therefore, there is some overlap between the content of this section and that presented in Chapter 4 - the reader is referred to the latter for more details on the various input and output controls that underlie use rights.

Second, whatever the form of use rights adopted, the following policy issues arise in implementing the use rights.

- The various forms of use rights may operate at the scale of the individual fisher, but equally they could be allocated to fishers' organisations or fishing communities; in other words, rights can be organised individually or collectively.

- The manner by which rights are distributed initially and over time may be driven by market forces, by a process of strategic planning, or by some other governing force.
- In implementing use rights, it must be decided what individuals or groups should be allowed to hold the rights, how the rights should be allocated initially, what should be their duration, and whether rights should be transferable from one user to another.

For each of these choices, decisions between the alternatives are critical from both a policy and a practical perspective; this matter will be explored in detail within section 5 of this chapter.

4.1 Territorial Use Rights

Among the most important management tools are those dealing with fishing location; these can be considered as arising in two forms. One is the 'closed area' approach, in which an entire fleet is affected equally by a blanket prohibition on fishing in certain locations (Chapter 3). The other, of interest here, is a rights-based approach involving Territorial Use Rights in Fishing (TURFs) and Customary Marine Tenure (CMT). These systems assign rights to individuals and/or groups to fish in certain locations, generally, although not necessarily, based on long-standing tradition ('customary usage').

A classic reference on TURFs is that of Christy (1982), who noted that "As more and more study is given to the culture and organisation of fishing communities, there are indications that some forms of TURFs are more pervasive than previously thought to be the case, in both modern and traditional marine fisheries." This point is echoed by others, who note the long-time and continuing operation of 'traditional sea tenure systems' around the world, and suggest that these systems hold considerable potential to provide relatively stable socially-supported fishery management.

Examples of TURFs are widespread – some examples include lagoon fisheries in the Ivory Coast, beach seine net fisheries along the West African coast, collection of shellfish and seaweed on a coastal village basis in South Korea and Japan, and controls over outsiders by fishing communities in Sri Lanka. Two particularly well-known examples are the long-standing arrangement in coastal Japan, where traditional institutions are incorporated in modern resource management, and the lobster fisheries on the north-eastern coast of North America, where fishers have been able to maintain extra-legal control on entry - exclusion rights. TURFs have a particularly long history in traditional, small-scale/artisanal and indigenous fisheries. Consider two examples, one in the Atlantic region of Canada, where the Mi'Kmaq (aboriginal) people have developed a social process for determining control over fishing territory, and the other in the artisanal fisheries of Chile:

> "In the centuries before the arrival of the first Europeans, the Mi'Kmaq... governed themselves through councils based on consensus in accordance with the laws of nature. District Chiefs were responsible... for confirming and reassigning hunting/harvesting territories." (Native Council of Nova Scotia, 1994).

> "Today, artisanal fisheries management measures in Chile consider the allocation of Territorial Use Rights in Fisheries (TURFs) among fishing communities traditionally exploiting benthonic resources such as Chilean abalone (*Concholepas concholepas*), sea urchins (*Loxechinus albus*) and macha clams (*Mesodesma donacium*), among others. Chilean fisheries legislation, the General Fisheries and Aquaculture Law (GFAL) enacted in 1991 allows the establishment of areas especially reserved for the use of specific artisanal fishing communities, through their legally constituted organizations (e.g., artisanal fishermens' associations and fishermens' cooperatives, among others)." (Gonzalel, 1996)

A common ingredient in these TURF systems is the local solution of usage issues. For example, Brownstein and Tremblay (1994) reported on the case of a small community in Nova Scotia, Canada faced with a lobster poaching problem in the late 1800s. The problem was resolved by the local church Minister, who decreed marine use rights based on an extension of property lines out to sea. In addition, if a fisher was unable to obtain a reasonable harvest from his or her area in a given year, the fisher would be given temporary access to a fishing 'commons', a reserve area designed to enhance equity in the fishery. Notably, this management system has proven workable and has been maintained by the community to this day.

Despite the many examples of CMT and TURFs, and the potential value of such systems both in their present form and adapted to other fisheries, such systems are generally poorly understood. As with any management mechanism, CMT and TURFs are not suitable in all cases, but depending on the specifics of the fishery, they may provide an efficient means of fishery management. For example, while some TURFs may involve excessive costs for development and maintenance, to make the arrangement work, others may be easily implemented and regulated within the framework of existing social institutions. Even if costly, the costs may be out-weighed by the inherent value of the institution involved. The point is that these options need to be examined and compared with the alternatives, particularly since, as Rowena Lawson (1984) has noted, such territorial management can be "the most effective method of control" especially if it can be "supervised by the fishing community itself or by its elected leaders." (See also Chapter 7.)

It is notable that, while many CMT and TURF systems have declined over time, there are now moves to maintain or re-establish some of these systems. For example, in the fisheries of Oceania, traditional CMT/TURF systems declined as fisheries commercialised, but initiatives in some nations (e.g., Solomon Islands, Fiji and Samoa) seek to re-establish them. Consider the case of Fiji. As noted by Veitayaki (1998), traditionally the principal marine resource management practice has been "the ownership of the customary fishing areas... by different, but closely related social groups" that regulate fishery use. The national government is seeking to reinforce this practice – in particular, it has "recorded, surveyed, and registered customary fishing grounds boundaries that, until now, were based solely on oral claims... and is planning to return to traditional communities the ownership of their traditional fishing grounds which at this time rests with the state...".

4.2 Limited Entry

Limited entry is a common management tool in which the government issues a limited number of licenses to fish (Chapter 4). This creates a use right - the right to participate in the fishery. Limited entry prevents the entry of new fishing boats and/or fishers, with the aim of controlling potential fishing effort (fleet capacity). If limited entry is successful, this limit on effort helps to conserve the resource and also generates higher incomes for the license holders (i.e., those holding the use right). Not surprisingly, this can be expected to be popular with those who actually obtain the use rights, while being opposed by others.

Limited entry has shown reasonable success in a variety of fishery management cases. For example, in a limited entry program for the Alaskan (USA) fishery, the state government reduced the number of active licenses, and license values in many fisheries remained high, indicating a relatively profitable fishery. On the Pacific coast of Costa Rica, the shrimp resources within the Gulf of Nicoya were exhibiting a decline in the 1980s, as the fishing fleet expanded. The introduction of a limited entry licensing program (and a corresponding ban on new boats in the areas) halted the growth in boat numbers; this move, combined with closed seasons, gear restrictions and other measures, led to some improvement in the state of the fishery system (Charles and Herrera, 1994).

The latter example illustrates a key message about limited entry, namely that it cannot be expected to 'solve' all management problems. Limited entry helps to prevent 'outsiders' from taking part in the fishery, but does not deal with managing the existing fleet. In particular, there remains an incentive for each fisher to try to catch the fish first, before the competitors get it (the 'race for the fish') – this, in turn, creates an incentive to over-expand vessel catching power (including both physical capacity and technology) beyond what is needed. As a result, there can be wasteful over-investment in the fleet, resulting in excess capacity, and pressure to over-exploit stocks. Therefore, limited entry, while a reasonable mechanism to assign use rights, must be implemented as part of a 'management portfolio' that also includes use rights applying to the existing fleet, such as rights over allocation of fishing effort or allowable catches (see below). Among the goals of this overall management portfolio should be the prevention or elimination of excess fishing capacity (Code of Conduct, Paragraph 7.1.8). The FAO International Plan of Action (IPOA) for the Management of Fishing Capacity[3] requires that States achieve 'world-wide, preferably by 2003 but not later than 2005, an efficient, equitable and transparent management of fishing capacity'. The overall goal of the IPOA is the reduction of fleet capacity world-wide to a level at which it no longer undermines long-term sustainable use of fishery resources.

It should be noted as well that the likelihood of success with limited entry will be much greater if it is put in place <u>before</u> the catching power of the fleet (or the number of participants) in the fishery becomes too large. In cases where this has not occurred, it has proven difficult to effectively reduce the number of licenses once the capacity is already excessive in relation to the productivity of the resource. Limiting entry is still important in such circumstances, but it becomes more of a challenge to bring the catching power in line with desired levels.

4.3 Effort Rights (Quantitative Input Rights)

Suppose that the catching power or capacity of a fleet, if unregulated, is greater than the fish stock can withstand. This implies the need to regulate how that catching power is used in practice, i.e., to limit the total fishing effort on the stocks so as to "ensure that levels of fishing effort are commensurate with the sustainable use of fishery resources..." (Code of Conduct, Paragraph 7.1.8). This can be approached in part through a limited entry scheme, to control the number of vessels fishing, but in addition, it may be desired to limit the amount of fishing by each fisher (or vessel), for example to allow more vessels to fish, for social reasons. To do this, the possible inputs that could be controlled include time fished, vessel size, amount of gear, and gear attributes. Such inputs might be controlled on an aggregate basis for the entire fishery or fishing fleet, e.g. by setting a total allowable number of boat days at sea for the fleet, or by prescribing allowable levels of inputs applying to individual fishers (such as a specific amount of fishing time and gear). The latter approach is one based on individual (input) use rights. (See Chapter 4 for further discussion on this topic.)

A common example of such an effort-based use rights approach arises in trap fisheries, notably those for lobster, crab and other invertebrates, where each fisher has the right to set a specified number of traps. It may be that all fishers have equal quantitative rights (i.e. to the same number of traps), or it may be that the rights vary from one individual to another, perhaps based on location, boat size, or some other criteria. Similar input rights are used with respect to fishing time at sea (e.g., in the USA) and vessel capacity. For example, Malaysia's "vessel replacement procedure" is such that if the owner of a fishing vessel wishes to build a replacement, the 'right' applies only to a new vessel of no greater size than the older one (FAO, 1998).

The key problem for an effort rights program is the incentive that will exist among fishers to thwart the input controls, by locating other uncontrolled inputs to expand (Chapter 4). This implies the need for a multi-dimensional approach to input rights, by implementing rights over

---

[3] The full text of the Plan of Action is available at http://www.fao.org/fi/ipa/capace.asp.

not one but a range of inputs. For example, in the lobster fishery of Atlantic Canada, access rights (limited entry) are supplemented with quantitative rights limiting the number of lobster pots per fisher - a relatively effective control for many decades. In recent years, however, changes in trap design as well as more frequent hauling and baiting of the traps have improved the effectiveness of each trap. Thus actual effort is not being held constant, which could lead to over-exploitation. To restore the effectiveness of effort controls, the set of use rights might be broadened to cover not only limited entry and trap limits, but also another dimension, the number of 'trap hauls'. In a similar vein, effective effort rights in a trawl fishery might cover a combination of hold tonnage, vessel horsepower and days fished.

Use rights over fishing effort must also deal with the natural process of technological improvement that gradually increases the effectiveness of any given set of inputs over time. If there is no compensation for this effect, conservation impacts of fishing by a given fleet may be under-estimated, leading to over-exploitation. However, it is possible for an individual input rights program to adjust for improvements in fishing efficiency, either by scaling back the rights over time (to reflect the rate of efficiency increase) or by placing the onus on vessel owners to ensure, and demonstrate, that efficiency increases (e.g., from addition of gear or from construction of new, presumably more efficient vessels) are compensated for by adjustments elsewhere in the fleet, so that overall catching power does not increase. Note as well that if fishers hold effective management rights, reductions in the level of input rights may take place voluntarily, on a collective basis – for example, with fishers in a trap fishery reducing the maximum number of traps allowed per fisher, both for conservation and cost reduction – as has happened in lobster fisheries in parts of Atlantic Canada.

Thus input/effort allocations can be a viable approach to rights-based management if care is taken in defining the rights, if a suitable portfolio of rights is established (cf. Hilborn et al. 2001), and if a plan is put in place to deal with fishing efficiency improvements and capacity control – as noted in the Code of Conduct (Paragraph 7.6.3). Note, however, that any quantitative rights system, whether involving effort rights or harvest quotas (see below) inherently requires certain data collection and monitoring schemes to operate; naturally, the cost and feasibility of these must be taken into account.

4.4 Harvest Quotas (Quantitative Output Rights)

A Total Allowable Catch (TAC) is a conservation control but not a use right, since setting a TAC makes no statement about the rights to catch the fish. The situation changes, however, if that TAC is subdivided into quotas allocated to sectors of the fishery, individual fishers, or communities, in which case these shares of the TAC represent quantitative output rights - collective or individual use rights over the corresponding 'shares'. Several variations on this may occur.

- The right may be held collectively by a sector of the fishery, with allocations made for small boats or large boats, for hook-and-line fishers or net fishers, etc., through a suitable institution within that sector.

- Rights may be assigned to communities, as 'community quotas', so that fishers within the community regulate themselves, perhaps with the involvement of their community, establishing suitable fishery management plans and dividing up the quota to suit their local situation and to maximise benefits, explicitly reflecting community values and objectives (Charles, 2001).

- Harvest rights may be allocated to individual fishers as trip limits (providing the right to take a certain catch on each fishing trip) combined with a right to a certain total number of trips per year (thereby ensuring that the TAC is not exceeded).

- Harvest rights may be allocated to individual fishers on an annual basis as <u>individual quotas</u> (IQs), rights to harvest annually a certain portion of the fish resource (a fraction of the TAC). These appear in two main forms: individual transferable quotas (ITQs) are harvest rights that can be permanently bought and sold among fishers in a 'quota market', while individual non-transferable quotas (INTQs) are rights that are not permanently transferable. (The impact of transferability is discussed later in this chapter.)

In contrast to effort/input rights, which have received relatively little research attention or promotion within fishery management, individual quotas have been studied and promoted heavily, particularly by fishery economists and participants in industrial fisheries. For example, a wealth of literature exists on ITQ systems in the best known locations, New Zealand and Iceland, as well as Australia, Canada and the United States. Individual quotas nevertheless remain rare in developing country fisheries, due to their considerable financial and personnel requirements (discussed below). Exceptions occur in some industrial fisheries, as with an INTQ system in Namibia, and ITQs in Chile, Peru and South Africa.

If individual harvest rights are considered certain within a fishing season, the fisher can plan fishing activity as desired, which can (a) potentially provide a better match to available markets, and (b) avoid the 'race for the fish', so that individual harvests can be taken at a lower cost, with less incentive for the over-capitalisation that can occur with limited entry and input allocation programs. This benefit exists both with trip limits and individual quotas, but more so with the latter since the fisher can plan fishing activity over the course of a full year, rather than on a trip-by-trip basis. IQ advocates argue that the above incentives lead to (1) reduction in fishery inputs such as fleet size and number of fishers, in keeping with the Code of Conduct (Paragraph 7.6.3), (2) increased rents in the fishery, and (3) increased product value, either through more attention to quality or through development of higher-valued product forms (e.g., fresh fish instead of frozen). Some evidence has emerged supporting these claims, although there is also some questioning of the extent of the benefits (e.g. Squires and Kirkley, 1996).

Potential benefits of individual harvest rights are accompanied with various possible social and conservation concerns. Social considerations are discussed later in the chapter, while conservation implications include those related to catch controls in general, and those arising due to particular incentives that exist to thwart individual controls (analogous to those discussed above for effort rights). Among the most widely discussed examples of these are the following.

- Inherent incentives exist for fishers to under-report catches, since every unreported fish is one less deduction from the fisher's own quota (and thus one more for catching later in the year) or one less fish for which quota must be purchased, at considerable cost, from other fishers. (This contrast with a competitive fishery, in which there is much less disincentive to report catches.)

- Incentives increase to dump, discard and high-grade fish, since this allows the fishers to directly increase the value of what he or she lands, thereby maximising profits obtained from the corresponding quota.

- In transferable rights systems (including some limited entry licensing approaches), fishers may go far into debt to purchase what may be expensive rights (quota) from others. This can lead to financial pressure on fishers to increase incomes and pay off the debt, and a resulting pressure to increase the TAC that may be difficult for decision-makers to resist.

In addition to conservation and social concerns, the same caveat noted for quantitative input (effort) rights applies here as well, namely that concerning data and monitoring requirements. These can be especially extensive in quota fisheries, since not only is it necessary to monitor the catches of individual fishers, the entire system is based on setting a sustainable Total Allowable Catch annually, which typically requires great scientific resources. (Indeed, such TACs have frequently been seriously mis-calculated even in cases for which financial and staff resources have been large.)

5. HOW ARE USE RIGHTS IMPLEMENTED?

The preceding discussion has outlined the range of possible use rights options. Here, we turn to the fundamental issue of putting a use rights system into place. This involves addressing the following three key questions.

- Is a system of use rights already in place within the given fishery? (5.1)
- If not, what use rights options, or set of options, are best for the given fishery? (5.2)
- What policy guides how the desired option(s) should be implemented? (5.3)

5.1 Are use rights already in place?

In existing fisheries, particularly those with a long history, it is crucial to understand whether use rights have already developed naturally over time, perhaps put in place by fishers themselves or by their community. This has proven to be the case in a wide variety of fisheries around the world. It is not surprising that use rights would have emerged, since there are clear benefits to defining the group of fishers entitled to fish in a certain locations, both for the fishers themselves and for the well-being of the fishing community. Social scientists have played a major role in documenting not only existing 'indigenous' use rights systems, but also systems that had been in place in the past, but which were displaced by 'modern' central management. In many cases, the process of understanding local use rights accompanies that of accessing local knowledge about the fishery and its environment – so-called traditional ecological knowledge (TEK).

If use rights already exist, the manager's first job may be to develop an understanding of how effective those use rights are, and whether there are available mechanisms to reinforce them. Certainly, it is likely to be more efficient to accept and reinforce existing rights rather than to attempt the development and enforcement of an entirely new regime. Thus, if use rights are currently in place, then only if they are for some reason unsustainable will it be necessary to explore how to implement a new use rights system.

5.2 What is the 'best' set of use rights?

For the remainder of the discussion in this section, we will assume that either no use rights system is in place, or that if use rights do exist, there is a recognised need for substantial changes. Then fishery management is faced with a choice among the many use rights options described in section 4. How do we compare output/harvest rights, input/effort rights, and TURFs (territorial use rights)? Can any one of these provide the 'best' solution? Several factors must be recognised in this regard.

1. Given the biological, economic and social diversity of fisheries, no single use rights approach will be applicable everywhere.

2. Each use rights option has its inherent advantages and limitations, and these will be of varying relevance depending on the specific fishery. Thus what is 'best' will depend on the fishery in question, and it is important to understand how the particular fishery circumstances influence the desirability of certain options over others.

3. Given the above two points, it is unlikely that any single use rights approach will produce optimal results. Thus it may be preferable to pursue a 'portfolio' of rights - a combination that is most acceptable, helps the fishery operate best, and maximises benefits for the given context.

These points highlight the reality that there is no single answer to the question 'what is the best use rights arrangement?'. We must be sceptical of any claim that any single use rights option is somehow inherently superior to others. Instead, fishery managers and planners, together with interested parties (such as vessel owners, crew members, aspiring fishers, community members and citizens), need to seek out or enhance a set of use rights that will work in practice. To this end, it is important to understand the structure and underlying nature of the fishery:

- what are society's objectives in the fishery;
- what are the relevant structure, history and traditions of the fishery;
- what is the relevant social, cultural and economic environment of the fishery;
- what are the key features of the fish stocks and ecosystem?

While there is neither a clear set of conclusions nor a consensus about which use rights options are most compatible with which fishery features, consider some possible ways in which responses to the above questions might guide our choices...

- Management of sedentary fishery resources may be especially amenable to territorial rights (TURFs).
- Management of stocks for which biomass estimates are unreliable, or for which regular catch monitoring is too expensive, may be best approached through effort rights rather than harvest rights.
- Management of highly migratory or transboundary stocks, for which the allowable catch must be allocated among nations, may focus on harvest rights. Management of fisheries in which the fishing technology is relatively uniform may focus on effort rights, while in fisheries with many different gear types, harvest rights may work better.

Of course, in a given case, the importance of each of the fishery characteristics must be weighed in assessing the pro's and con's of use rights options, before arriving at a desired solution.

5.3 What is the underlying policy framework?

Whether the goal is to enhance and reinforce existing use rights arrangements, or to develop a new set of use rights (as discussed above), we arrive at the matter of determining the specific framework for implementing use rights. Whatever the use rights system being pursued, there are several policy issues that arise concerning the allocation and governance of the rights. How precisely should the desired use rights option(s) be implemented? How are those rights to be 'managed'? What management institutions will be effective for the various combinations of fishery resources, industry structure and political jurisdictions? Who should be involved in establishing and operating a use rights system? Suitable policies are needed to guide these decisions.

It is important to recognise that the matter of use rights is likely to be a controversial and delicate one. After all, use rights define who can and cannot take part in the fishery. In addition, there is likely to be an element of irreversibility to any decision about use rights; once rights are allocated, it may be very difficult to make major changes. The task of implementing use rights will be made easier if clear policy directions are laid out in advance, since such policy should provide guidance in terms of which fishery interested parties are to receive priority in obtaining use rights (e.g., small versus large vessels, community versus corporate participants, etc.). Finally, it is also important to keep in mind that decisions involving use rights can affect not only current fishers but potential participants as well. This implies that despite the broad usefulness of co-management arrangements, in which current fishers take part in management, it may be considered unfair to restrict participation in discussions of use rights to current fishers. The question of who should be allowed to take part in such discussions (and indeed whether a participatory process is feasible at all) needs careful consideration (see Chapter 7).

The sensitive nature of use rights, described above, is particularly relevant when the *status quo* set of use rights is seen as inappropriate in the context of national policy directions. This may arise when a nation is undergoing a major transformation, for whatever reason, as is the case, for example, with some countries in eastern Europe, Central America and southern Africa. In South Africa, for example, the transformation from a period of apartheid into one of democracy means that broadening the right to access the fishery (as with other parts of the economy) is a matter of urgency (Cochrane and Payne, 1998). In such circumstances, national policy goals drive use rights decisions within the fishery sector.

5.3.1 Should use rights be governed by market forces or strategic planning?

A key issue in fishery policy debate over use rights concerns the mechanism by which the holding of rights is itself managed. In many cases, this revolves around the choice between two institutional arrangements for determining who are to be the fishery participants – a market-based approach, and one based on multi-objective planning, often at the community level.

A reliance on market forces has become a popular direction among many governments and international financial institutions. This leads to a market-based approach to fishery policy, as typified by ITQ systems, in which strategic-level issues over use rights - who is to participate in the fishery and who is to receive allocations of allowable catch or effort - are determined through the buying and selling of rights in the marketplace.

Who it is that buys, or sells, the rights will depend on the situation at hand. It may be, as economic theory suggests, that more-efficient stakeholders buy out less-efficient ones, or it could be that the buyers are those with better access to financial capital (a particularly important point in many developing countries), or there may be some other factor that dominates. Furthermore, while market-based rights are typically discussed in the context of individual fishers, there is nothing conceptually to prevent an entity operating at a collective (corporate or community) level from buying or selling on a fishing rights market. However, the actual bundle of rights resulting from such a transaction may differ depending on whether the buyer is an individual, corporation or community - due for example to differing regulatory constraints on the various rights-holders.

Broadly speaking, a market-based rights system can be expected to display the various advantages and disadvantages of the overall market system, and inspire similar debates to those arising with respect to market mechanisms elsewhere in the economy. For example, depending on one's perspective, and the case at hand, markets may (or may not) be the most cost-efficient institutional arrangement to handle transactions between fishers, and may (or may not) increase flexibility in fisher operations. Given a widespread familiarity with markets in many economies, these may be relatively easily-implemented fishery rights systems – but may introduce financial impacts on the pursuit of new policy directions (since, for example, those with market-based rights must be compensated if policies are contrary to the self-interest of rights-holders).

In contrast, a strategic planning approach assigns use rights in a more deliberate manner (whether permanently or periodically) through a decision-making process that (a) is based on recognising multiple societal goals, (b) is carried out by institutions operating at a suitable scale, whether community-based, regional or national, and (c) involves rights specified through a combination of legislation and government decisions, on the one hand, and traditional/informal arrangements on the other. These rights may operate at the individual fisher level, but (as with the market approach) could alternatively be allocated at the group (collective) level, with allocations made through relevant institutions – see below. Such arrangements have a lengthy history in real-world situations – arising frequently, for example, in the context of cooperatives, marketing boards, and indigenous/native communities.

5.3.2 Should use rights be individual rights or community-based rights?

One of the most crucial aspects to consider about use rights is the difference between those instituted at the level of the individual fisher and those at a collective level – e.g., the community or the fishers association. In one fishery, the government fishery authority may have specified a certain group of licensed fishers, and designated each individual's right to fish a certain amount of gear. Thus use rights – a license and an effort right - are at an individual level. In another fishery, use rights may be held by a coastal community or a fishers' association, which then determines which individuals will take part in the fishery at a given time. In such a case, rights may be primarily at the group level, but may be allocated down to an individual level. (Note that relative to individual rights, collective rights are typically less well understood, so particular attention is devoted to such rights in this section.)

The choice between individual and collective rights should depend on both the historical context and the fishery objectives being pursued. For example, in the case of a fishery that has developed relatively recently, and that has an industrial focus, there may be a natural inclination to an individual rights system. Individual rights are often seen as naturally compatible with the entrepreneurial independence of fishers. On the other hand, collective rights are historically of greatest importance in longstanding traditional fisheries – although such rights have not always been properly understood and incorporated into 'modern' management, leading in some cases to severe social and conservation problems.

Collective rights cannot be expected to work in every fishery. It is a challenging question, and indeed a matter of some research attention, to determine conditions conducive to introducing these rights, but it seems likely that factors needing consideration include: the cohesiveness of the community involved, experience in and capacity for local management, geographical clarity of the community, and its overall size and extent (see Section 8, Chapter 7 for further discussion on this topic).

For those cases where such rights already exist, or where the conditions are conducive to their introduction, collective rights have the potential to provide considerable benefits, notably in fisheries for which the community has a strong inherent interest. Through moral pressure on local fishers and by providing suitable management institutions, the community can create a collective incentive for resource stewardship (conservation) as well as increased management efficiency, and implementation of local enforcement tools.

Consider the example of 'community quotas' - fishing quotas (portions of the TAC) allocated to communities rather than to individuals or companies. While suffering some of the inherent flaws of any quota-based scheme, community quotas defined on a geographical basis tend to bring people together in a common purpose, rather than focusing on individualism. Fishers in the community manage themselves, perhaps also with the involvement of their community. The fishers create fishery management plans (Chapter 9) and divide up the quota (or other form of rights), to suit their specific local situation and to maximise overall benefits, rather than leaving it to the market to make choices for them. Note that this involves use rights at both the community level and the individual level. As well as providing many of the benefits of individual quotas, this approach may also enhance community sustainability, allowing each community to decide for itself how to utilise its quota. For example, one community may decide to allocate its quota in a rent-maximising auction, while another may prefer distributing the quota so as to achieve a mix of social objectives, such as community stability, employment and equity. Examples of this approach include Alaska's system of Community Development Quotas (CDQs) and the fixed-gear community management boards in the Scotia-Fundy groundfish fishery of Atlantic Canada.

5.3.3 What should be the duration of use rights?

Within any system of use rights, it is a fundamental matter to decide how long the rights are to last, i.e. for how much time the holders of rights are able to make use of those particular rights.

This revolves largely around the balancing of two factors: management flexibility and conservationist incentives. On the one hand, short-duration rights give the capability to more frequently re-allocate those rights, a flexibility that may allow fishery management to better reflect society's changing objectives over time. On the other hand, longer-duration rights, by providing more security to fishery users, give those users a stake in the well-being of the resource further into the future, and an incentive to better 'plan for the future' in husbanding the resource.

Consider two examples. First, suppose that to operate in a particular fishery, a substantial level of financial investment is needed, and this is held by only a small set of industrial companies. Since these are the only interested parties able to fish, they may be given long-duration use rights. However, if an objective of management is to improve the situation of artisanal fishers, this goal may be thwarted for many years; such fishers may be unable to enter the fishery even if they develop the financial means to do so, e.g. through cooperatives. Shorter-duration rights might have provided management flexibility to facilitate access of artisanal fishers much earlier. (This situation could arise as well if a foreign fleet has dominated in a particular fishery, but the national policy goal is to develop a domestic presence. In such a case, allocating long-duration rights to the foreign fleet may be counter-productive.) Alternatively, suppose that short-duration rights are allocated in a fishery. What happens as the end of the time period approaches? If the rights-holders know that their rights are about to expire, they may have an incentive to harvest intensively, with no regard for the future of the fish stocks. Rights of longer duration (a longer 'time horizon') would mean that incentives to conserve would apply for many more years than would otherwise be the case.

There is no universal 'right answer' in dealing with such trade-offs, and indeed it is important to note that the trade-offs themselves are not as stark as described above. Flexibility can be added to systems of long-duration rights (e.g., by allowing targeted transferability of rights), while incentives to avoid over-exploitation can be provided in the case of short-duration rights (e.g., conservation performance criteria placed on the option to renew the rights).

In many fisheries, use rights tend to be of indefinite duration. This is particularly the case in small-scale/artisanal fisheries, where access rights may be available to all those in the local community, and those rights may be considered essentially permanent. This may be desired if no particular need exists for the flexibility to re-allocate rights, and the idea of limiting access rights to a certain period of time is deemed unacceptable. In particular, it seems unlikely in such a context that a government manager would force a local fisher to leave the fishery after, say, 5 years of fishing, on the basis that the person's time had 'run out'. (However just such an occurrence could be contemplated in the same fishery, if it were managed on a community basis – in some tribal fisheries, for example, the community accesses the resource collectively, and there may be utility in alternating those who actually do the fishing for the community.)

Explicit limitations on the duration of use rights seem more common in commercial or industrial fisheries, in which leases or other agreements may allow harvesting over a limited number of years. This is especially common with coastal states establishing access rights for foreign fleets, on an annual or multi-year basis, but an explicit duration also arises in some domestic fisheries. Examples include the auctioning of periodic harvest rights for inland fisheries in Bangladesh, and Namibia's individual non-transferrable quota system, in which "the term of a right can be 4, 7 or 10 years" depending on the company's levels of investment and of Namibian ownership (Oelofsen 1999). As alluded to above, it is crucial in such situations to develop mechanisms that reduce the incentive users may have to over-exploit the resources when the term of the use right nears an end.

It is clear from the discussion above, and the range of possibilities described, that there is a close linkage between decisions about the duration of rights and the matter of deciding who is to hold the use rights. We turn now to a discussion of the latter issue.

5.3.4 Who should hold use rights?

For any given fishery, some people hold use rights and others do not. In a tribal fishery, it may be the Chief who decides who is to have access to the resource. In a fishery with limited-entry licensing, the government's fisheries authority may designate license holders. Whatever the situation, the issue arises: who <u>should</u> hold use rights? This is largely a policy-level decision, i.e. one that will reflect overall policy in the fishery and beyond. Some guidance is provided, for example, by the Code of Conduct, which notes (Paragraph 6.18): "States should appropriately protect the rights of fishers and fishworkers, particularly those engaged in subsistence, small-scale and artisanal fisheries, to a secure and just livelihood, as well as preferential access, where appropriate, to traditional fishing grounds and resources in the waters under national jurisdiction."

Whatever the chosen form of use rights in a given fishery system, two key issues arise in implementing the use rights.

- Firstly, who should hold the use rights initially, and how should the corresponding initial allocation of rights be carried out?

- Secondly, who should hold use rights in the future, and in particular, should fishery users be permitted to transfer (buy, sell and trade) rights among themselves?

These questions are discussed in the following two sub-sections.

5.3.5 How should use rights be allocated initially?

If a use rights system already exists in a fishery, the distribution of the rights among the various participants (and the exclusion of non-participants) has been already established. But what if a new use rights system is being implemented, or there are to be adjustments to the existing system? Then a critical step in implementation is to determine how to allocate/distribute the rights. To put it another way: who is to receive which rights? This is a highly contentious part of the process, and it is important to realise that there is no universally 'correct' way to allocate rights. Perhaps the challenge involved here can be summarised as that of dividing up the total set of rights in a manner that seeks to minimise conflict. This process must be accompanied by some form of appeals process to manage the special cases that arise. While such processes sometimes resolve conflicts over initial allocations of rights, the debates involved at this stage can be so extensive as to delay implementation of rights systems for years.

Among the allocation mechanisms that might be considered are the following.

Auctions. The theory developed for quantitative fishing rights notes that to maximise economic efficiency, it may be desirable to auction the rights. With such an approach, those willing or able to bid highest will acquire the rights, regardless of historical, social or cultural considerations. Because of this lack of social sensitivity, this approach is often seen as unpalatable, and is little used - although it may be useful in purely industrial fisheries or situations where social factors are considered irrelevant. One example of its use has been the allocation of inland fishery rights in Bangladesh.

Catch History. Political reality in many cases is such that use rights have been allocated to fishers on the basis of the historical participation of those involved. This is usually referred to as a 'catch history' approach, since it most often focuses on allocating rights in proportion to each individual's past catches. This can be problematic, however. What is the best way to define historical participation? Consider the case of individual catch quotas. If only recent catch histories are used in the allocation, then fishers who were temporarily not fishing are penalised. This sends a questionable signal, especially if the period over which the 'history' is calculated was one of over-fishing and stock depletion. Those who receive the lowest quotas are those who contributed least to the over-fishing. On the other hand, those who bent the rules (or fished illegally without being caught) – perhaps using fine-mesh gear to increase catches, or dumping less desirable fish to fill their holds with higher-valued species - are rewarded, into perpetuity, with a larger quota. While such a situation is unfair, the alternative of avoiding recent history will lead to objections from recent entrants, who may be technologically advanced and politically powerful. In any case, there is a strong possibility that social conflict will arise.

History + Equity. In an effort to deal with such problems, hybrid schemes for initial allocations may be used. For example, there is some experience with formulas in which part of the total is allocated based on catch history and the remainder is allocated equally among fishers.

Allocation Panel/Board. Since the auction mechanism for allocating use rights is inherently affected by differences in levels of wealth among prospective participants, and the catch history approach is dependent on previous performance in the fishery, neither is particularly suitable when public policy involves broadening the base of involvement in the fishery, to include those who were previously excluded. (A dramatic examples of this would be the case of fisheries in South Africa; see Cochrane and Payne, 1998.) In such cases, a special body may be required to determine allocations of rights to specific groups or communities.

Community/Sector/Group Allocations. Note that among the above methods, auctions tend to directly involve individuals (or companies), a catch history approach can be at an individual or collective level, and panels/boards may be best suited to the allocation of use rights at the level of community, sector or grouping. In some cases, it may be advantageous to use a combination of approaches in a 2-step process. Rights could be allocated initially solely on a collective basis, directly to communities, fishing sectors or other identifiable groups. The second step in the process is then devolved to each community or grouping, involving the determination of exactly which individuals are to obtain rights. Of course, this does not eliminate the challenge of how to allocate to the communities or groups, nor does it eliminate problems at the individual level, but for the latter, it does facilitate solutions that are tailored locally. For example, a community may consider it important not just to allocate rights amongst the existing local fishers, but to balance their interests with crew members as well as others who would like to participate, but have been excluded in the past, or have just now reached an age or seniority at which participation is possible.

5.3.6 Should use rights be transferable?

Once use rights are allocated, there remains the key question of whether or not to make rights transferable. In other words, can the rights be bought and sold, or handed down in a family from one generation to the next, or temporarily transferred to another fisher within a fishing season? This question is closely related to the debate discussed above concerning the extent to which the market will govern who holds fishery rights. Several approaches are possible:

1. *completely non-transferable* rights can only be used by the holder, and are no longer valid when that fisher leaves the fishery;

2. *non-divisible transfer* of use rights (whether fishing licenses, input allocations or quota rights) may be allowed between fishers, but only if done as a complete indivisible package - i.e. with all a fisher's use rights transferred together;

3. *divisible transfer* of use rights is the ultimate version of uncontrolled transferability, in which fishers can freely sell all or any portion of their rights;

4. transferability may be allowed only within the fishery sector or community in which the use rights reside, for options 2 and 3 above, thereby providing greater fishery stability within the sector or community; or

5. a hybrid approach could be adopted in which there are differing classes of licenses, some transferable and some not, with defined policy measures determining which fishers have

which form of licenses (for example, there could be two classes of license: transferable ones for full-time fishers, and non-transferable ones for part-timers).

These options have very different implications depending on whether we are considering temporary or permanent transferability. Consider first the case of temporary transferability, in which one fisher is permitted to rent or lease use rights to another fisher within a fishing season. The rights then revert to the original fisher at the end of the season. This mechanism provides important flexibility so that a fisher who happens to become sick or whose vessel breaks down one year can still obtain some income by renting out the use rights. As long as regulations prevent the excessive use of this mechanism, there would seem to be few long-term impacts. On the other hand, with permanent transferability, the implications are more significant. The remainder of this section focuses on this case of permanent transferability, exploring several key issues, relating to efficiency, fishery mobility, social cohesion and concentration of rights.

Efficiency. Transferability is often promoted as a means to improve economic efficiency, using an argument such as the following. To be economically efficient, the participants in a fishing fleet should be those most profitable in harvesting the available fish. In theory, a market-based system (such as an ITQ system), with divisibility and transferability of input or output rights, improves efficiency, as vessel owners who maximise the profits resulting from a given quota will buy up that quota from others - like a commodity on the market. The idea is that with transferability, the more 'efficient' vessel owners remain in the fishery, while others sell their quota and leave, in a 'survival of the fittest' process leading to increasing overall efficiency of individual fishers.

This seems a persuasive argument, but it is useful to highlight some serious potential qualifications to the argument, which lead to the overall conclusion that an appeal to increased efficiency does not universally support transferability, or non-transferability - the outcome will depend on our specific objectives and the specific fishery. First, there is no guarantee that efficiency will increase through the market process. For example, if there is strategic buying of quota to gain greater control of the fishery (not unlike the manner by which financial mergers take place), ownership will still become concentrated and participation reduced in the fishery, but the resulting impacts on efficiency are unclear.

Second, while efficiency - obtaining optimal benefits for a given set of inputs, or 'doing the most with what we have' - is desirable, it must be assessed at a level appropriate to the policy goals. Specifically, while transferability may increase the efficiency of individual vessels (with higher vessel profits), it could actually decrease efficiency for the fishery as a whole, as well as the coastal economy and coastal community. This is because when we consider the bigger picture, we must take into account (a) all stakeholders rather than just the owner of an individual vessel, and (b) all related monetary and non-monetary benefits from catching a fish, not only profits to the boat owner. These related benefits will depend on the specific situation, but would typically include benefits to crew members and on-shore workers, as well as to the on-shore economy and relevant coastal communities. Such considerations are often ignored in looking at the economics of the fishery, but must be taken into account if we are to properly assess the desirability of use rights transferability.

Third, not only must efficiency be seen broadly, in the context of the whole fishery and the coastal economy, it must also be seen as a long-term conservation matter. This has a variety of implications. For example, if use rights are transferred out of a local area, thereby reducing the use of locally-based traditional ecological knowledge in management, we need to be conscious of possible negative conservation impacts (Chapter 7). Also, since we need to regulate the impact on fish stocks, an 'efficient' fishery should be seen as one that produces the greatest net benefits for every fish caught. This implies that it is not a matter of getting large quantities of fish quickly and cheaply out of the sea, but rather getting the most from each fish that is taken. There is no reason to expect that buying and selling of transferable rights will reflect this broader idea of efficiency.

Fisher Mobility. Transferability increases the 'mobility' of individual fishers, allows each to exit the fishery when the revenue to be gained from the sale of the use right exceeds the expected benefits of remaining in the fishery. This provides maximum flexibility for the fisher, and makes it easier for managers to reduce participation in the fishery. However, in the absence of restrictions to keep use rights within the local community, this mobility can reduce stability within the fishing communities. Conversely, non-transferrable systems provide better stability, but reduce mobility of the fishers - making it more difficult to reduce fishing power over time (capacity reduction). In particular, incentives exist to keep non-transferable rights in use as long as possible, to maximise actual benefits, and in the hope of a financial windfall should there be a later decision to allow transferability. This may mean that a boat will be used beyond its technological life, which can also create safety problems.

Social Cohesion. Transferability can have a major impact on social well-being. First, since rights are often held solely by the vessel owner, the selling of those rights may leave the most vulnerable in the fishery – crew members – without jobs and without compensation. Second, transferability can lead to a loss of social cohesion in the community as a whole. This is particularly the case when transfers remove fishing rights from the community, resulting in reduced local involvement in the fishery, reduced employment, and a corresponding increase in the proportion of 'outsiders' fishing on what had been locally-controlled resources. On the other hand, non-transferrable rights may help stabilise the local economy, by ensuring that a certain portion of the rights resides in the local communities. It should be noted, however, that in the case of non-transferrable systems, there may be inherent pressure to shift to transferability, with all the implications of that change - it seems that this has occurred in most examples to date.

Concentration of rights. Transferability generally leads to the concentration of fishing rights ownership in fewer hands (and particularly in the hands of processors or dealers). If the goal is to reduce the number of stakeholders with which the manager needs to interact, transferable rights may accomplish the goal. On the other hand, concentration of rights raises social and economic concerns, particularly in relation to potentially detrimental impacts on:

- the traditional organisational arrangements of fishers;
- employment of vessel crew;
- overall equity in the coastal economy;
- the stability of fishing communities.

To prevent concentration, rights might be made non-transferable, or there might be limits on the maximum amount of rights that can be owned by any person or firm, or an owner-operator requirement might be put in place (so only the vessel owner can be its operator). However, there are means to evade these approaches – e.g., through legal contracts or nominal ownership (by family, relatives, and employees). Thus, if concentration of fishing rights is considered undesirable, caution is needed, since even supposedly non-transferable rights might end up being transferable in practice.

6. SYNTHESIS

Many of the major ongoing debates in fishery management relate to use rights – the right to use (access) a fishery – as well as <u>management rights</u>, the right to take part in its management. After discussing the relationship between use rights, management rights and other forms of fishery rights, this chapter presented the range of use rights options, including TURFs, limited entry and individual quantitative input rights (effort) and output rights (quotas). The majority of these use rights options correspond with specific input and output controls, as described in Chapter 4. Finally, the chapter reviewed a range of policy and operational issues relevant to the choice of use rights, and to their implementation. It should be clear that use rights can be highly controversial, due not only to the conflict inherent in any use rights system (where some are excluded, and some have numerically greater rights than others) but also to the lack of consensus on the 'best' path to follow in establishing such systems.

On the latter point, suppose that we are involved in management of a specific fishery, and we are certain that no (acceptable) use rights system is currently in place. Then we are faced with the challenge of developing and implementing such a system. What policy approaches are most suitable? Consider, for example, the choices described above, (a) between a market-based and a strategic planning mechanism to govern use rights, and (b) between an individual and a collective or community-based orientation for the use rights. There is no consensus on precisely which characteristics of a fishery favour one or the other of these two approaches. However, Berkes (1986) has suggested, in the context of Turkish fishery cases, that community-based strategic planning "provides a relevant and feasible set of institutional arrangements for managing some coastal fisheries", particularly "small-scale fisheries in which the community of users is relatively homogeneous and the group size relatively small". On the other hand, he suggests that individual market-based use rights (perhaps including the "assignment of exclusive and transferable fishing rights") may be appropriate "for offshore fish resources and larger-scale, more mobile fishing fleets". Building on these suggestions and the discussions in the earlier parts of this chapter, we might envision a 'working hypothesis' as follows:

Community-based Rights if:	Market-based Rights if:
- structure is small-scale/artisanal with clear fisher-community ties	- the fishery has a predominantly industrial, capital-intensive orientation
- history and tradition play a major role in fishing activity and fishery management	- the fishery does not play a major role in supporting coastal communities
- multiple fishery and non-fishery goals are pursued; fishery management requires the balancing of these objectives	- profitability dominates over community and socioeconomic goals (e.g. equity, employment, health of local economy)

This decision framework should not be seen as a recommended one, but merely as a possibility. Even if it has some validity, there are bound to be exceptions, and it is not clear what circumstances will lead to such exceptions. Furthermore, the 'either/or' nature of the above – a completely collective rights systems and a totally market-driven system – fails to incorporate the wide range of intermediate options.

Consider but one example, that of non-transferrable rights. Fisheries in Namibia (Oelofsen, 1999) provide an illustration of how transferability of individual rights (e.g., ITQs) is "not regarded as the ideal system to be implemented" because it would make more difficult the pursuit of national policy goals. Similarly, long-duration rights ("the notion of fixed rights in perpetuity") are seen as unsuitable, since the positive qualities of a fishing company that may have led to it being given a quota could change over time. In Namibian fisheries, therefore, limited-term individual non-transferrable rights (INTQs) were implemented as an intermediate approach.

In certain cases, non-transferrable rights could help to balance the benefits of individual rights with the goals of social and community stability. For example, such a system of use rights might prohibit permanent transfers/sales, but allow the flexibility of rights transfers within a season.

Furthermore, any individual rights might operate subject to community rights and locally-set rules, with fishers in the community having group control over the system of rights.

In considering the range of use rights options, it is important to note that the 'community-based versus market-based' dichotomy must be differentiated from that arising between input/effort and output/catch rights. Unfortunately, these two issues tend to become confused, probably due to the domination of output-based ITQs within the market-based approach. However, we must keep in mind that various combinations, such as community quotas or market-oriented effort controls, may be feasible, depending on the specific fishery context.

The discussion in this final section has focused on just one of the range of policy issues surrounding use rights in fisheries. There is clearly much more on this complex subject than can fit in one chapter. Thus, we have sought here only to introduce the concept of fishery rights, the range of options, and some of the key operational and policy issues. Some key points to be considered in assessing and developing use rights systems are highlighted in Figure 2.

Perhaps four fundamental points should be reiterated in conclusion:

- use rights are crucial in the pursuit of responsible fisheries;
- use rights already exist in many fisheries;
- if use rights do not exist, or current rights are ineffective, an appropriate system must be developed and implemented;
- this task requires considerable care, with no simple 'cook-book' formulas to help.

These realities ensure that use rights will continue to play a major role in fishery management. Indeed, it seems apparent that if both the use rights held by fishers and the responsibilities undertaken by those fishers are clearly identified and widely accepted, success in achieving responsible fisheries will be that much more likely in the future.

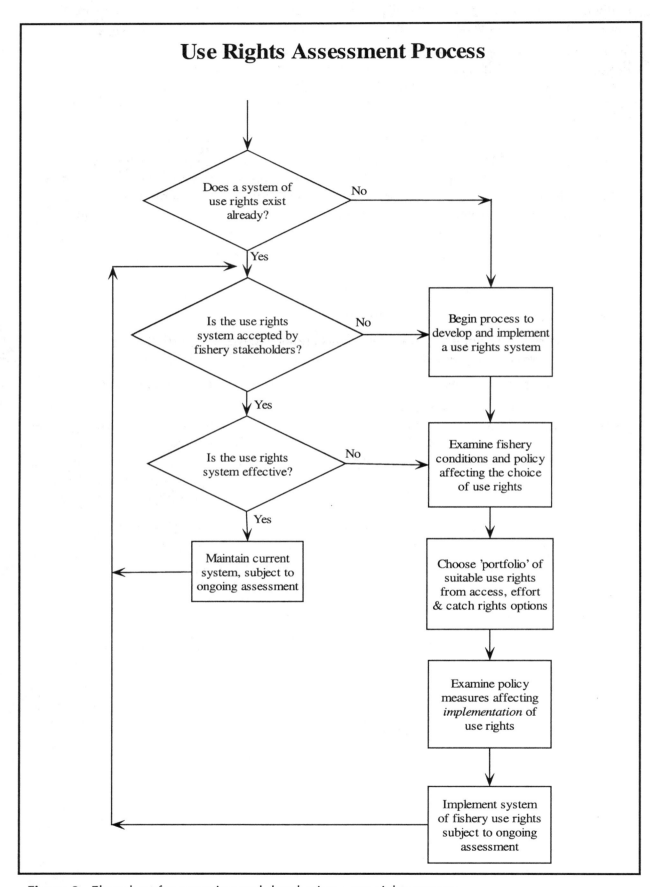

Figure 2. Flowchart for assessing and developing a use rights system.

7. REFERENCES AND RECOMMENDED READING

Berkes, F. 1986. Local-level management and the commons problem: A comparative study of Turkish coastal fisheries. *Marine Policy*, **10**: 215-229.

Brownstein, J. and Tremblay, J. 1994. Traditional property rights and cooperative management in the Canadian lobster fishery. *The Lobster Newsletter*, 7:5.

Charles, A.T. 1998a. Living with Uncertainty in Fisheries: Analytical Methods, Management Priorities and the Canadian Groundfishery Experience. *Fisheries Research*, **37**: 37-50.

Charles, A.T. 1998b. *Fisheries in Transition*. Ocean Yearbook 13. Eds. E.M. Borgese, A. Chircop, M. McConnell & J.R. Morgan. University of Chicago Press, Chicago, U.S.A. p.15-37.

Charles, A.T. 2001. *Sustainable Fishery Systems*. Blackwell Science, Oxford, United Kingdom.

Charles, A.T. and Herrera, A. 1994. Development and diversification: Sustainability strategies for a Costa Rican fishing cooperative. *Proceedings of the 6th Conference of the International Institute for Fisheries Economics and Trade*. Eds. M. Antona, J. Catanzano, & J.G. Sutinen. IIFET/ORSTOM, Paris, France.

Christy, F.T. 1982. *Territorial Use Rights in Marine Fisheries: Definitions and Conditions*. FAO Fisheries Technical Paper, **227**. FAO, Rome, Italy.

Cochrane, K. L. and Payne, A.I.L. 1998. People, purses and power: the role of policy in directing fisheries management as indicated in the debate surrounding a developing fisheries policy for South Africa. In Pitcher, T. J., Hart, P. J. B. and Pauly, D. eds. *Reinventing Fisheries Management*. (Proceedings of Symposium held in Vancouver, Canada, February 20-24, 1996). Kluwer, Dordrecht. p. 73-99.

Congress of the Philippines. 1998. *An Act providing for the Development, Management, and Conservation of the Fisheries and Aquatic Resources, Integrating all Laws Pertinent thereto, and for other Purposes*. Republic Act No. 8550. Congress of the Philippines, Republic of the Philippines. Metro Manila, Philippines.

Copes, P. 1997. Social impacts of fisheries management regimes based on individual quotas. *Proceedings of the Workshop on Social Implications of Quota Systems in Fisheries*. Vestman Islands, Iceland, May 1996. Nordic Council of Ministers, Copenhagen, Denmark.

Dewees, C.M. 1998. Effects of individual quota systems on New Zealand and British Columbia fisheries. *Ecological Applications*, **8** Supplement.: S133-S138.

Dyer, C.L., and McGoodwin, J.R. 1994. *Folk Management in the World's Fisheries: Lessons for Modern Fisheries Management*. University Press of Colorado, Niwot, U.S.A.

FAO. 1998. *Report of the FAO Technical Working Group on the Management of Fishing Capacity. La Jolla, United States of America, 15-18 April 1998*. FAO Fisheries Report, **586**. FAO, Rome, Italy.

Gimbel, K.L. (Editor). 1994. *Limiting Access to Marine Fisheries: Keeping the Focus on Conservation*. Center for Marine Conservation and World Wildlife Fund. Washington, U.S.A.

Gonzalel, E. 1996. Territorial use rights in Chilean fisheries. *Marine Resource Economics*, **11**: 211-218.

Hanna, S. and Munasinghe, M. (eds). 1995a. *Property Rights and the Environment: Social and Ecological Issues.* Beijer International Institute of Ecological Economics and the World Bank. Washington, U.S.A.

Hanna, S. and Munasinghe, M. (eds). 1995b. *Property Rights in a Social and Ecological Context: Case Studies and Design Applications.* Beijer International Institute of Ecological Economics and the World Bank, Washington, U.S.A.

Hardin, G. 1968. The tragedy of the commons. *Science,* **162**: 1243-47.

Hilborn, R., Maguire, J.-J., Parma, A.M., and Rosenberg, A.A. 2001. The Precautionary Approach and risk management: Can they increase the probability of success of fishery management? *Canadian Journal of Fisheries and Aquatic Sciences,* **58**: 99-107.

International Institute for Rural Reconstruction (IIRR). 1998. *Participatory Methods in Community-Based Coastal Resource Management.* International Institute for Rural Reconstruction, Silang, Philippines.

Lawson, R. 1984. *Economics of Fisheries Development.* Frances Pinter Publishers, London, U.K. 283pp.

Native Council of Nova Scotia. 1994. *Mi'kmaq Fisheries Netukulimk: Towards a Better Understanding.* Native Council of Nova Scotia, Truro, Canada.

Neher, P.A., Arnason, R. and Mollett, N. (eds). 1989. *Rights Based Fishing.* Kluwer Academic Publishers, Dordrecht, Netherlands.

OECD. 1993. *The Use of Individual Quotas in Fisheries Management.* Organisation for Economic Co-operation and Development. Paris, France.

Oelofsen, B.W. 1999. Fisheries management: the Namibian approach. *ICES Journal of Marine Science,* **56**: 999-1004.

Ostrom, E., and Schlager, E. 1996. The formation of property rights. In: *Rights to Nature: Ecological, Economic, Cultural and Political Principles of Institutions for the Environment.* Eds. S. Hanna, C. Folke, & K.G. Mäler. Island Press, Washington, U.S.A. p.127-156

Pinkerton, E., and Weinstein, M. 1995. *Fisheries that work: Sustainability through Community-based Management.* The David Suzuki Foundation, Vancouver, Canada. 199pp.

Ruddle, K. 1989. The continuity of traditional management practices: The case of Japanese coastal fisheries. In: *Traditional Marine Resource Management in the Pacific Basin: An Anthology.* Eds. K. Ruddle & R.E. Johannes. Contending with Global Change Study No.2, UNESCO/ ROSTSEA, Jakarta, Indonesia. p.263-285

Ruddle, K., Hviding, E. and Johannes, R.E. 1992. Marine resources management in the context of customary tenure. *Marine Resource Economics,* **7**: 249-73.

Squires, D. and Kirkley, J. 1996. Individual transferable quotas in a multiproduct common property industry. *Canadian Journal of Economics,* **29**: 318-42.

Symes, D. (ed.). 1998. *Property Rights and Regulatory Systems in Fisheries.* Fishing News Books (Blackwell Science), Oxford, United Kingdom.

Symes, D. (ed.). 1999. *Alternative Management Systems for Fisheries.* Fishing News Books (Blackwell Science), Oxford, United Kingdom.

Townsend, R.E. 1990. Entry restrictions in the fishery: A survey of the evidence. *Land Economics,* **66**: 359-78.

Townsend, R.E. and Charles, A.T. 1997. User rights in fishing. In: *Northwest Atlantic Groundfish: Perspectives on a Fishery Collapse.* Eds. J. Boreman, B.S. Nakashima, J.A. Wilson & R.L. Kendall. American Fisheries Society, Bethesda, U.S.A. p. 177-84

Veitayaki, J. 1998. Traditional and community-based marine resources management system in Fiji: An evolving integrated process. *Coastal Management,* **26**: 47-60.

CHAPTER 7

PARTNERSHIPS IN MANAGEMENT

by

Evelyn PINKERTON

School of Resource and Environmental Management, Simon Fraser University,

British Columbia, Canada

1.	INTRODUCTION: PARTNERSHIPS SOLVE PROBLEMS, BUT ARE LITTLE KNOWN BY MANAGERS	159
2.	PARTNERSHIPS OF SMALL AND LARGE SCOPE	161
3.	PARTNERSHIPS OF SMALL AND LARGE SCALE	163
4.	PARTNERSHIPS WITH DUAL OR MULTIPLE PARTIES	165
5.	PARTNERSHIPS WITH DIFFERENT LEVELS OF COMMUNITY EMPOWERMENT: ACCOUNTABILITY	165
6.	UNUSUAL PARTNERSHIPS SOLVING PARTICULAR EQUITY PROBLEMS: LINKING OFFSHORE FISHERIES TO COASTAL COMMUNITIES	166
7.	POWER DIFFERENTIALS OF DIVERSE ACTORS ON REGIONAL BOARDS	168
8.	CONDITIONS FOR EFFECTIVE PARTNERSHIPS	169
	8.1 Characteristics of the partners	169
	8.2 Characteristics of the partnership or the institution created through the partnership	170
	8.3 Characteristics of the resource(s)	171
9.	COMMUNITY PARTNERS MAY ADD VALUE TO THE RESOURCE	171
10.	CONCLUSION	171
11.	ACKNOWLEDGEMENTS	172
12.	REFERENCES	172

1. INTRODUCTION: PARTNERSHIPS SOLVE PROBLEMS, BUT ARE LITTLE KNOWN BY MANAGERS

Chapter 1 outlined the global problems facing fisheries managers, especially overfishing and habitat loss, and the related local problems of poor data and data analysis, the low legitimacy of regulations and the consequent poor enforcement of regulations. This chapter outlines how new forms of partnering between local communities and/or non-governmental organizations (NGOs) and government regulators can address many of these global and local problems, and analyzes examples of such partnerships. These examples, and many others in the literature, demonstrate that users can be usefully involved in *all* the functions and responsibilities of fisheries management identified in Figure 1 of Chapter 1, from information gathering to analysis to plan formulation to rule-making to enforcement and compliance. The notion that users can and often should be involved at all stages of management, rather than consulted by government regulators as an "add on" when fishing plans and regulations are mostly complete, can involve a substantial change in the conception of management, if decision-making is currently highly centralized.

This chapter addresses not only the classic partnerships between government and small fishing communities, but also more complex and innovative ones between parties operating at different scales. For example, small communities and large fishing enterprises can be involved in co-management agreements together. Examples of multiple-scale partnerships are rarer but equally important for solving problems, especially where there are interactions between offshore and inshore fisheries.

But if partnerships are so useful, one might ask, why don't we have them already? The short answer is that partnerships have to be carefully designed to be appropriate for the situation, as well as accountable and effective. Not all situations are ripe for partnerships, nor will partnerships be successful under all conditions. The good news is that we now have some good success indicators. In the last 20 years social scientists have assembled a rich body of documentation which allows us to identify the conditions and situations which are good predictors of successful partnerships. This chapter summarizes some of this literature.

Unfortunately, knowledge of the existence of partnerships, as well as what it takes for them to succeed, has been largely absent from the training and experience of fisheries managers, as well as from the management agencies in which natural scientists work. Despite this, many natural scientists have recognized the need to integrate "human dimensions" into our management systems. Peter Larkin reminded us a decade ago that in focusing on fish, we often forget that we manage fish mainly through managing people. As we shall see in this chapter, managing people means understanding how human organizations and human values can work towards, rather than against, the goals of fisheries management.

Indeed, our failure to manage people effectively has turned out to be the main limiting factor in fisheries management today. No matter how well-designed fisheries harvest regulations or habitat protection measures appear to be, if fishers or polluters can find a way around them, management efforts are frustrated. But when communities or organizations of fishers are included as partners in the planning, design, and implementation of the regulations, when they participate in protecting habitat, and even more, when they are part of the crafting of the very policies which underlie management decisions, they grant full legitimacy to the regulations, and are the strongest advocates, monitors, enforcers, and implementers of management decisions. Community/NGO partners may even help agencies reconceptualize a problem and develop a better strategy for attacking it. A recognition of the importance of these kinds of partnership roles in fisheries management is reflected in Paragraphs 6.13 and 6.16 of the Code of Conduct for Responsible Fisheries.

Involving communities or NGOs in management may also be the only way that enough resources can be mobilized to manage effectively. In some cases, this means resources to make even basic stock abundance assessments. In other cases, it means resources to consider species interactions and ecosystem values in management. Our understanding of natural systems has evolved to the point that we know we need to develop management systems which reflect the complexity and diversity of what is being managed (Chapter 9). Yet we lack the flexible governance structures and resources to achieve this. We know governments will never be able to achieve this ambitious task alone. A growing literature shows how important aspects of this challenge are already being met through partnerships.

Partnerships vary in the scope of their activities, in the geographic scale of the marine or freshwater areas involved, in the types and number of parties involved, and in the degree of power held by the non-governmental parties. The next few sections explore examples of different types of partnerships along these four spectrums, and the costs of developing them.

2. PARTNERSHIPS OF SMALL AND LARGE SCOPE

At the smallest scope, a government manager may partner with a local group to collect data on a local stock or sub-stock, or monitor pollution, or habitat loss. Much of the most exciting recent research on local or traditional ecological knowledge is based on this type of arrangement (Neis and Felt, 2000). Local fishers may already have extensive knowledge of the behaviour and abundance of a local stock. Managers may work with local fishers simply to record their knowledge and annual observations of changes; or they may actively incorporate the fishers into the work of controlled monitoring of the stock (see Paragraph 6.4 in the Code of Conduct). Often such work provides the only information government has on a local fish population. Since many species are made up of smaller distinct local substocks, such information may be the most critical data informing management of the entire species.

Alternatively, knowledge of the substock may reveal a spatial or temporal pattern which is masked by survey data taken on a larger scale and within a different time frame. Local monitoring of lobster on the Magdalen Islands provided species-wide information on their spatial distribution, use of habitat, and interactions with other species, data which was heretofore missing (Gendron et al., 2000). Local fishers usually have developed hypotheses about these relationships based on decades of observation. They often suggest factors which should be considered, often ones which would not occur to managers.

In summary, the benefits of such a small-scope partnership can be multiple and unobtainable without a partnership with local fishers.

The partnership may:

- provide reliable data on the abundance and composition of a local stock or sub-stock (which might be used as an indicator stock, if a species is composed of many stocks);

- help managers interpret large-scale changes in abundance and the environment;

- suggest hypotheses about relationships which scientists can then test;

- assist scientists with the most effective sampling techniques and sites for ongoing monitoring or research.

Suppose that, in addition to collecting and interpreting data, the fishers also assisted in interpreting the effect of regulations on fishing behaviour, and even assisted in crafting more effective regulations. Thus local knowledge would be used to analyze regulatory failures and to proactively design more successful approaches. This was done on Magdelen lobster, where local fishers helped managers interpret the meaning of catch data, helping them understand that fishing efficiency had increased despite tighter regulations. This led managers to consider that reducing fishing effort through time and area closures might be a less effective method of reducing mortality than, for example, reducing licences or traps (Gendron et al., 2000).

Partnering in this slightly larger scope of management:

- checks the validity of managers' interpretation of catch data;

- suggests alternative methods of regulating catch.

The foregoing types of small-scope partnerships can be very low-cost if fishers are willing to volunteer time and the use of pre-existing equipment, and if government can provide some start-up staff time to develop a relationship with fishers who become identified as reliable monitors.

Volunteerism is likely in situations which are already part of a work day or pleasurable activity for fishers, or where the fisher is learning something new and interesting, or is being made part of an activity which offers greater access to information or power. However, long-term willingness to volunteer is likely to persist only if managers are able to recognize and treat their fisher partners with respect. Ideally managers will perceive that many fishers have a passionate interest in the health of the resource and have much to offer as interpreters of data if they are given access to the data and an opportunity to reflect on its possible meanings. A valuable relationship of mutual respect and trust is likely to occur if the most knowledgeable and reflective fishers are treated as colleagues whose opinions are valued. In such a relationship, they would be given open access to government data and an opportunity to participate in both data analysis and the design of regulations based on the data, activities which are discussed in some detail in Chapter 5.

In many cases, NGOs can contribute small amounts of staff time or funding to support a key monitoring activity. Volunteerism and NGO donations are more likely to be forthcoming if the management agency is forthcoming about its own budget situation, and/or is able to contribute some modest financial support of its own. To do this, managers often need to overcome deep-seated attitudes that fishers are incapable of acting in anything but their own short-term, narrow self-interest. They need to remember that fishers will often act in the long-term and broader public interest if they see that they also will benefit in the long run. In other words, fishers and communities will invest in their future if they have assurances that they will be part of the benefits to accrue from present restraint or volunteer effort. (A more in-depth discussion of the conditions which produce this occurs below).

Small-scope projects that are low cost are most effective if they can make a tight connection between the sense of or identification with place which often exists in fishing communities and the objectives of the monitoring or other small-scope activity. For example, better data on a particular local stock will enable more sensitive and appropriate fishing plans on that stock and a better understanding of what other factors affect its abundance in the local area. Regulations based on such improved data will be considered more legitimate and will be better obeyed, and fishers are likely to continue volunteer monitoring, if they are assured they will have the first access to recovered stocks (see also Chapter 6).

Partnerships of the largest scope are those which include fishing communities or NGOs in all aspects of management, even the identification of policy issues and the making of policy. A highly developed example of a large-scope partnership is the system in Washington State, U.S.A., in which the treaty tribes co-manage the salmon fishery with the Washington Department of Fish and Game. The tribes meet with the state governor's council alongside all the departments managing natural resources, to identify policy issues of concern and to develop approaches. In other words, the tribes have a government-to-government relationship with the Department of Fish and Game, even though they do not have equal resources. This type of partnership is explored below under "different levels of power." In the evolution of their relationship, the tribes and the state have developed transparency of data arrangements through the University of Washington computer, and divided up management tasks to avoid duplication.

Large-scope partnerships normally involve extensive staffing and capacity-building of the NGO partners, and are normally funded through multiple government agencies. In the United States of America, the tribal managers are funded largely through the Bureau of Indian Affairs. Canada, however, is experimenting with funding large-scope regional organizations through the Department of Fisheries and Oceans. (This usually happens after long-term conflicts result in the building of cooperation, and the Department believes management will be better served by funding regions on an experimental basis, institutionalizing them in the long term if they are successful). These organizations have accessed funding from private foundations and multiple government departments in their start-up phases, however. The research initiatives of these organizations are also attracting interest from universities, as well as government researchers, and can often become hubs of funding and activity from multiple private and public sources. The advantages of large-scope partnership can be summarized as follows.

- Small-scope partnerships may attract volunteerism, contributions-in-kind, and small cash donations, while requiring only part-time staff attention from government.

- Large-scope partnerships need substantial external funding sources, but may attract at least some funding from foundations or NGOs, due to their innovative nature and their capacity to conduct integrated research on ecosystem linkages. They may eventually be self-funding if they become non-profit membership-based organizations.

- The use of volunteer labour and contributions in kind (use of vessels and equipment) will be sustainably offered if a relationship of trust and mutual respect can be established.

- Data transparency and an openness to varying interpretations of data are highly desirable as a mechanism for building trust and mutual aid.

It is important to note that in many developing countries and island states, there are fishing communities that informally regulate their fishing effort, based on their observations of fish abundance and their reading of indicators which predict fish abundance over time. These cases have been documented, summarized and synthesized by many geographers, anthropologists, political scientists, and biologists, including Johannes (1978, 1981), Klee (1980), Spoehr (1980), Ruddle and Johannes (1985), Cordell (1989), Dyer and McGoodwin (1994), Wilson et al. (1994), Schlager and Ostrom (1993), Durrenberger and King (2000). Japan has achieved the most "complete" and integrated set of partnerships, in that it has integrated ancient local systems of management into fisheries planning at all levels of local, regional, and national government (Yamamoto and Short 1992, Pinkerton and Weinstein 1995). A reading of these cases suggests that an important challenge for developing countries may be to identify whether such home grown systems already exist and, if so, to support and integrate them into regional and national management through co-management agreements. Funding to support such systems' start-up phase is sometimes accessible through international research organizations or universities (Stoffle et al. 1994).

3. PARTNERSHIPS OF SMALL AND LARGE SCALE

The foregoing examples of partnerships of small scope were also ones of small scale, since the geographic area is small (but note that temporal scale could be extensive in a small area). Ames et al. (2000) point to the need to incorporate multiple spatial and temporal scales in the analysis of how fishing alters ecological processes. This aspiration is contrasted with the conventional approach to management decisions, based on measuring the fishing mortality of a single stock at a single spatial scale–the range of the fish–and at a single temporal scale–one year. Scientists have long noted that overfishing is often a function of unrecognized habitat degradation. Of course, the need to protect fish habitat will not be addressed unless managers understand how to monitor habitat and species linkages. Thus analysts argue that overfishing is often undetected unless managers monitor the species composition of the ecological community to which the fish belong and the gradual removal of patches or subsystems in a progressive manner. The management implication of this argument is that local groups may be the best placed to protect the small-scale components of an ecosystem. Thus the management actions of local fishers or local partners should be integrated into an adaptive multi-scaled governing institution to match the multiple scales of the fisheries system (Ames et al., 2000).

Some of the most successful partnerships function at a regional scale, around a common interest in a watershed or sub-basin. A small-scope regional-scale partnership, as exemplified in the Kuskokwim River in Alaska, is data collection on the same stocks by different communities located at different points on the river as the stocks swim upstream. By combining their data on a real-time basis, as well as their traditional local knowledge of the migration patterns of different stocks under different conditions, the various communities on the river are able to work with the Alaska Department of Fish and Game to produce a good assessment of stocks and species abundance. This is effective because a regional-scale partnership can cover a larger range of conditions and factors than the single test fishery or counting fence which was formerly operated by government. The partnership arose out of controversy over the accuracy of government's abundance index. With the regional-scale partnership, the abundance estimate is now considered legitimate, and government has a far richer data base to inform its decisions (Pinkerton and Weinstein, 1995). The partnership also was based largely on volunteer monitoring by fishers, with donations in kind by a regional association, a federal government agency, and a fish processing company. The Alaska Department of Fish and Game contributed staff time during the first season to train a community-based monitor, who thereafter became a highly-trusted partner, and staff time for in-season meetings to compile and interpret the data in collaboration with community-based partners.

A larger-scope regional-scale partnership might factor basin-wide habitat affects and enhancement activities into its fishing plans, as did the Skeena Watershed Committee in British Columbia, or the Mitchell River Watershed in Australia. This means that multiple community parties and multiple government agencies make decisions based on multiple criteria. From the community perspective, fishers more readily accept curtailment of opportunity to take less abundant species if they participate in the planning process to increase the abundance of other species and the planning for the restoration of the depressed species. The planning in this case includes improving habitat protection, engaging in habitat restoration, and engaging in activities to enhance the productivity of freshwater production of fry and smolts. Perhaps most important is that the involved fishers and community representatives on the committee see a broader picture than their own narrow self-interest, and feel part of a grander scheme to restore the health of the watershed. Being able to get beyond geographic isolation and cooperate on a regional scale can have a powerful effect on the parties. This occurs because they perceive that by cooperating, they can have the power to effect positive change on a more meaningful scale, at the same time that they improve or at least stabilize their own position. They also perceive that improvements in the fishery will not occur unless all differently-situated parties contribute to the solution. Because of this, they are often willing to "give to get", even delaying their own "pay-off". Government agencies may be likewise enabled to get beyond turf battles and make trade-offs in the interest of getting a better outcome in the long run (Pinkerton and Weinstein, 1995).

The Skeena Watershed Committee process was part of an ambitious and expensive government experiment involving a great deal of stock research by government scientists, and the hiring of a professional facilitator. However, it also generated funding from other government agencies and private sources, as well as donations of time and effort by all participants. Depending on human resources available, it would not be impossible to develop regional-scale cooperation without major funding. In sum:

- small-scale large-scope partnerships may be part of the multi-scaled forms of management necessary to monitor and analyze the dynamics of progressive habitat loss and species interactions;

- regional-scale small-scope partnerships may combine multiple sources of data to create a more accurate real-time picture of stock abundance and the affect of fishing pressure;

- regional-scale large-scope partnerships may enable an analysis of fishing/habitat interactions;

- regional-scale large-scope partnerships may enable cooperation and greater pay-off flexibility among differently-situated sectors when they collaborate to increase fish abundance for their mutual benefit.

4. PARTNERSHIPS WITH DUAL OR MULTIPLE PARTIES

The simplest partnership is between one government agency and one community or NGO. But whatever the scope of management decisions shared, or the geographic scale of the management unit, partnerships often include multiple "communities" and multiple government agencies, as in the watershed management examples above.

Such partnerships are being modelled in Canada on both coasts. On the west coast of Vancouver Island in the province of British Columbia, a partnership between aboriginal First Nations and non-aboriginal communities, alongside municipal and regional governments, local environmental groups, and community development agencies has been developing since the early 1990s. It was formalized in 1997, the terms of reference were approved by February 2001 and formation of the board began in November 2001 (Pinkerton, 1999; Loucks et al., 2002). On the east coast of Canada, multiple gear groups and community representatives have formed boards along county lines to create fishing plans and to monitor fish deliveries (Loucks et al., 1998; deYoung et al., 1999). On both coasts, these local planning bodies are nested within broader regional organizations of multiple local parties and communities which are guided by a broad public interest.

The benefits of multiple communities being able to collaborate are enormous. From government's perspective, there is a conflict-resolution benefit, which often solves allocation problems between warring sectors: sport, commercial, and aboriginal sectors, whose struggles were damaging to good management. From the community/NGO perspective, collaboration means that the chances of developing and implementing a long-term vision and developing a sense of stewardship around that vision are greatly increased. The fact that the vision has input from multiple perspectives and that some of the partners are often communities (who have a general interest in the long-term health of the region, as opposed to fishers, who may have a vested interest in particular stocks) means that the vision has broad support and is likely to be sustainable. The importance of developing such partnerships is addressed in, for example, Paragraphs 6.13, 7.1.2 and 7.6.6 of the Code of Conduct.

5. PARTNERSHIPS WITH DIFFERENT LEVELS OF COMMUNITY EMPOWERMENT: ACCOUNTABILITY

The larger the scope of management activities in which a community or regional board is involved, the more likely it is that the level of power held by the NGO is high. However, since power-sharing is usually negotiated, it is possible that a community/NGO might hold significant power in one area of management, and little power in another. For example, an agreement to co-manage a fishery could involve equal power in developing a fishing plan, but no community/NGO power in deciding who had access to the fish, or no community/NGO power is making general policy about the direction or vision of future management goals.

In other words, there are different levels of power, or power over different levels of decision-making, whatever the scale of decisions. Equal partnership in deciding how to collect and analyze data is an important, but still a relatively weak, form of power compared to the power to decide how the fishery will be conducted or to decide who has the right to participate in a fishery, how much they get to take, etc. We might think of a hierarchy of levels of power. At the bottom of the hierarchy are decisions about operations or activities, which can themselves be arrayed in order of ascending importance. At the next level are decisions about who gets to make the operational rules, and who is excluded from the management area affected by the decisions (membership). At the top of the hierarchy are decisions about how the rules at the other two levels have to be made, and what groups may participate in making them (Ostrom, 1990). The treaty tribes in Washington State hold co-equal status with the Washington Department of Fisheries at each level of decision-making, and thus hold the maximum possible partnership power, as well as the maximum scope in decision-making (in all areas of management). (In contrast, the geographic scale of the territory over which each tribe has authority is relatively small, so the tribes coordinate their negotiations though a body with no legal authority, the Northwest Indian Fish Commission).

All of these levels of power can be contrasted with the mere advisory status that is often granted to fishers' associations by government managers (Berkes et al., 1991). It is important to recognize how advisors differ from partners, and that the advisory relationship does not garner the benefits of the partner relationship. If advisors perceive that they have little power and influence, they will revert to the usual divisive client politics. This points to the major benefits of power-sharing: conflict resolution and the mobilization of energy to solve problems in critical parts of the system. This applies equally to conflict resolution between government and communities and among competing parties, who may participate in regional co-management through regional partnerships.

It is often not appreciated by government that human resources are a valuable form of energy, which may be mobilized and utilized only under the right conditions. Sociologists tell us that people will "go all out", contributing far beyond what is formally required of them when they: (a) believe in the goal of a partnership, (b) feel they are part of a working team on which there is mutual respect and concern for the welfare of all partners, (c) feel that they are able to make a contribution to the team which is respected and honoured by the other partners (Senge, 1990).These three conditions apply in fisheries management partnerships where there is a situation of accountability between and among partners. Accountability requires:

- transparency of data;

- an ability to discuss differing interpretations of the data;

- an ability to agree on what the basic problems are and what approach is most promising;

- clear agreements to share decision-making;

- clear articulation of the standards being used to evaluate decisions and their results; and

- an ability to have timely feedback on outcomes of decisions.

When accountability is lacking, human resources are not mobilized. Furthermore, energy flows in the opposite direction. Instead of working overtime to solve problems, fishers and communities actively subvert management plans and actions by managers which they don't feel are accountable.

6. UNUSUAL PARTNERSHIPS SOLVING PARTICULAR EQUITY PROBLEMS: LINKING OFFSHORE FISHERIES TO COASTAL COMMUNITIES

In addition to accountability, one of the necessary conditions for successful partnership is equity. Equity normally refers to the democratic representation of different gear groups or differently-situated fishers or community representatives on co-management boards which share power with government.

Most nations also face another perplexing equity dilemma. How should they balance the operation of highly-capitalized and highly efficient offshore fleets with the access needs of coastal communities which support many small-scale artisanal or subsistence fisheries? This dilemma has been well-captured in a documentary film about the nation-wide strikes of inshore fishers and small-scale fish marketers in India in the 1990s. The government of India wanted the offshore trawl fleet as an important source of cash and foreign exchange. The inshore fishers and marketers, however, noted that species of fish which they used to take in their artisanal inshore fisheries were being wiped out. They staged national protests and eventually succeeded in having the offshore fleet shut down, or at least its non-Indian components (Thalenberg,1998).

In most cases, it is probably not realistic or even desirable to shut down the offshore fleets completely. In some cases these fleets may take species which are not available in inshore waters. However, it is often the case that these fleets not only depend on migratory species which are also taken inshore, but that they take as bycatch many other species on which inshore fishing communities are dependent. So there is usually a policy dilemma in how to balance these two needs. An innovative partnership in Alaska called the Community Development Quota (CDQ) program has been used to address important aspects of this dilemma.

In Alaska, about 10% of a billion dollar fishery on Bering Sea pollock, halibut, sablefish, crab, and other groundfish that had been taken mainly by offshore fleets (based in centres of production distant from Alaska), was allocated to six coalitions of villages (comprised of 62 villages in total) in Western Alaska. The villages, which had been traditionally dependent on inshore fisheries (but not pollock) were geographically isolated and had limited access to sources of cash income. The goal of the program, begun in 1992, was to help the communities to develop the infrastructure and have personnel necessary to support long-term participation in the industry, and thus build a stronger economic and social base (National Research Council, 1999). The program aimed to address the exclusion and marginalization of these communities from the industry, and from access to all fisheries. This was considered key, because even the original licences in salmon, herring, and halibut allocated to these villages tended to be sold into urban centres or larger communities elsewhere in Alaska. In some cases, the villagers had never received fishing licences, having fished only for subsistence.

The community coalitions are organized as non-profit corporations which set goals and objectives, and submitted annual business plans to the Alaska Department of Community and Regional Affairs. Reports on their performance in meeting these goals are reviewed by the state, which has exercised considerable oversight. The state has the authority to reallocate quota among the six corporations, based on their performance, and has already exercised this authority.

The communities do not have a direct role in fisheries management decisions in the pollock fishery (the major large-scale offshore fishery), but their presence in the fishery as partners or deckhands means that they are likely to assert their interest in the bycatch and/or habitat destruction by this fleet where it affects species they take in community-based fisheries (mainly halibut and salmon). And they do have a management role in the smaller-scale community fisheries on other species which most villages have created from the proceeds of the pollock fishery. The communities essentially receive a royalty from the industrial fleet, which they use to develop their own participation in the fishery or for education. These activities are overseen by the National Marine Fisheries Service, the federal agency which has jurisdiction outside 3 miles from the Alaska coast. The communities may: receive the royalty as cash; negotiate jobs for community members on board pollock vessels to which they lease their catch share; use the cash for scholarships, or to buy gear or vessels or licences; lease a quota share to community fishers; or create a local halibut skiff fishery.

Where communities created a local fishery, they used seasons and trip limits to spread opportunities among fishers. They used primarily skiffs up to 36 feet in length, and the CDQ organizations kept track of the harvest levels and controlled the pace of the fishery (Langdon, 1999). These new CDQ fisheries have not created a new "race for fish" nor overcapitalized vessels, but have remained largely small-scale. One village association has constructed a few larger vessels, but makes them fish five miles outside the village so that the local skiffs have a territory reserved for them. Some villages have added value and kept jobs in the community by constructing small processing plants. Some for-profit plants held by the non-profit organizations have also withheld 20% of fishers' landings to ensure that start-up loans are paid off in a timely manner. In short, the villages have had a vision of how to integrate a commercial fishery into their subsistence economies through a development plan (Langdon, 1999).

The Alaska CDQ program could be applied more generally to any fishing-dependent communities with limited economic opportunities. The community partners may have rights to make decisions only about their own membership, and how to conduct their own fisheries in inshore areas. Such partnerships can address difficult policy issues around equity, however. They illustrate how:

- partnerships can be used to create co-ventures between capital-intensive fleets and community-based fisheries which offer greater opportunity to communities otherwise forced out of the fishery by economic conditions and forces;

- partnerships can focus attention on the need to reduce bycatch by industry fleets of species that are also taken by inshore fisheries communities;

- partnerships between the state and fishing-dependent communities can be used to foster wise development of new community-based fisheries that are not overcapitalized, and that are within the means of the community to plan;

- community quotas can be used as mechanisms for allocating non-transferable fishing opportunities to communities; flexibility in allocation can be created, however, if the state can transfer quota based on performance of community-stated goals;

- community quotas are also a mechanism for dealing with the power differential between large and well-organized economic actors and small, dispersed economic actors which are nonetheless a large sector of the economy and the welfare of many nations

If it is recognized that this sector does not compete well for access rights to fisheries, yet plays a key role in the social and economic diversity and well-being of a country, then a country can use this mechanism to assure the continued role of communities in the fishery.

7. POWER DIFFERENTIALS OF DIVERSE ACTORS ON REGIONAL BOARDS

Another special case of equity involves the dilemma of how to represent powerful non-local interests on regional boards, where these external actors have fishing rights in the region that will be affected by regional board decisions. The problem is that external actors cannot be expected to have the same level of concern as does the community about protecting local habitats, ecosystem linkages, or the sustainable harvesting of local stocks. This is because external actors are less identified with the region, have less opportunity to develop a stewardship ethic, and are more likely to have other diversified opportunities (are less dependent on the local stocks and their habitats). They could be characterized as having an economic interest in the region, but not a stewardship interest. The West Coast Vancouver Island board is solving this problem by having such actors represented on committees including local fishers which develop fishing plans, but not represented directly on the board which makes overall policy. This leads to the following important corollary.

Condition: External economic interests can be represented on community or regional boards, as long as they do not have an opportunity to dominate them

Another effort to model the linking and creating of equity between offshore fleets (of large vessels) and nearshore or inshore fishing groups (using smaller vessels) is the Northwest Atlantic

Marine Alliance (NAMA), founded in 1997 in Saco, Maine, USA. NAMA is currently drawing support from fishers' organizations in the states of Maine, New Hampshire, Massachusetts and Rhode Island as well as Maritime Canada. NAMA is a non-profit umbrella coalition promoting collaborative research in order to provide education about ecosystem linkages and selective fishing. NAMA also facilitates efforts by stakeholders to craft regulations which do not disadvantage inshore vessels, or transfer fishing effort from offshore areas to inshore ones, thus threatening the historical allocation balance and geographic spread of fishing effort. Because fishing regulations by the New England Fishery Management Council can be insensitive to local conditions, NAMA promotes the rights of local areas to develop regulations for their local areas which are more appropriate than the generic ones made by the Council. Many of the latter tend to increase fishing effort by offshore fleets and increase bycatch and wastage through regulatory discards due, for example, to very low daily trip limits. NAMA advocates balance and communication between offshore and inshore within a vision of ecosystem-oriented conservation based on community linkages, increased awareness about fishing practices destructive to ecosystem values and linkages, and the need to enhance stocks. It seeks to include all relevant and affected marine resource interested parties in its membership (http://www.namanet.org ;interviews 2000). The existence of NAMA is another illustration of the fact that conservation initiatives within industry often emerge from the inshore, more community-based sector, but that this sector is capable of reaching out to and integrating the offshore sector under the right leadership and conditions. NAMA enjoys the support of high-profile public figures, and an ideology in the state that fishing-dependent coastal communities should not be the first to be pushed out of the industry. The non-profit organization is funded through foundation grants, individual and corporate contributions and membership. NAMA employs two staff and receives the equivalent of two more staff positions through the volunteer work of board members.

Condition: Organizations which increase communication and education among different sectors and gear types of the commercial fleet can promote a stewardship ethic in all sectors and increase cooperation and appropriate regulations of different sectors.

8. CONDITIONS FOR EFFECTIVE PARTNERSHIPS

So far I have discussed conditions enabling partnership design which affect accountability and equity. In this section I focus on what conditions are good predictors that a partnership will be effective. Some of these conditions pertain to the characteristics of the partners; some pertain to the characteristics of the partnership or the institutions created through the partnership; and some conditions pertain to the characteristics of the resource(s) being managed through the partnership. Social scientists are not able to state categorically which conditions are necessary in all circumstances, or in what combinations with each other, but only that the more conditions that are present in a particular case, the more likely it is that success can be predicted. It should be noted that these conditions for effectiveness need to be considered together with the conditions for accountability and equity discussed earlier.

8.1 Characteristics of the partners

- Communities or regions which have a **high level of dependency** on the resource have a greater incentive to develop sustainable use patterns, and are more oriented toward learning how to do this. This is because they are very vulnerable to non-sustainable use.

- Communities or regions which are **highly identified with their geographic area**, and are thus unwilling or unable to transfer access rights out of the area, are more likely to develop a stewardship ethic.

- Communities or regions whose **membership can be readily defined** have the potential power to exclude non-members and retain at least some of the benefits of management with the membership. This allows the development of incentives to invest in management.

- Communities or regions where **committed and credible leadership** exists, and where an energy centre or sparkplug is able to push the agenda forward, are more likely to sustain an effort to overcome the barriers to innovation.

- Communities or regions where a core group of people are **willing to invest enough time** in building the agreements to address the problems which are required to see the process of creating partnerships and then operating them.

- Communities or regions where some **homogeneity of values, customs, norms, activities** already exists, and/or where important sub-groupings have already built some degree of trust and understanding are part way down the road to partnership already. It is possible to build agreements from scratch, but far easier if some social capital already exists in the form of shared understandings.

- Communities or regions where sufficient **local knowledge** of the resource exists among people who are willing to share it offer far greater incentives to government to share power, and provide a basis from which to design more appropriate regulations.

- Communities or regions where **skills exist at building consensus or agreements** among community members will more readily mobilize their communities and build solid problem-solving local co-management bodies.

- A government agency which is **oriented toward learning** will more readily negotiate an adaptive and flexible agreement with partners.

- A government agency containing at least key individuals who have the **political will** to make the partnership happen, and who will work behind the scenes to overcome obstacles.

- A government agency with some **willingness to delegate or decentralize** enough decisions to make partnerships possible.

8.2 Characteristics of the partnership or the institution created through the partnership

- Permits **common access** to data and data analysis on the status of the resource.

- Permits the **making of appropriate regulations** for the local/regional situation (regarding both fishing and habitat protection).

- Permits the **monitoring** of compliance to these regulations by both government and community partners.

- Permits the **enforcement** of the regulations.

- Permits assurances that **investments will be rewarded**: improvements made in the resource by the management work of the fishing partner will benefit, at least partially, the partner who invested the resources to make the improvement.

- Permits the **resolution of conflict** in a timely manner, through informal or formal means, and with agreed appeals to a higher body if conflict is not resolved.

- Permits access to **sufficient start-up monetary resources**, where these are required by the scope of the partnership.

8.3 Characteristics of the resource(s)

- The resource is amenable to **boundaries defining resource management units.** In the case of migratory stocks, amenable to agreements being made horizontally or vertically with other regions sharing the management of the passing stock.

- The resource occupies/uses habitat/territory **adjacent to the community** or frequented by community members or fishers.

- The resource is **capable of being monitored** by community fishers or members.

- The resource is currently or potentially of **sufficient abundance and value** (or to supply some key ecosystem service) to be of interest to the community.

9. COMMUNITY PARTNERS MAY ADD VALUE TO THE RESOURCE

An increasing portion of the world's fish are harvested and processed using mass production strategies which miss opportunities to add value. That is, far more fish are sold in their lowest-value form than is required by market demand, simply because this fits the production strategy of large firms (Pinkerton, 1999).

Mass production at the fishing stage also encourages overfishing, because the capital intensive vessels require high volumes of fish to cover operating costs, especially considering market fluctuations in fuel and fish prices. Boom and bust market cycles force vessels with high operating costs to take more fish to cover costs, and fisheries they target are degraded (Clapp, 1998).

Community-based fisheries are smaller-scale, less capital intensive, less sensitive to changes in operating costs (labour, fuel), and hence more flexible in adapting to fluctuations in world fish prices, or to changes in fish abundance. Because they are smaller-scale and more labour-intensive, they have more opportunities–at least in the presence of appropriate preservation technology- to capture the fish live, or to preserve the quality of the fish longer, and to process it into a higher quality (more value added) product (deYoung et al., 1999).

10. CONCLUSION

This chapter has discussed the conditions under which the following problems may be addressed through partnerships: poor data and data analysis; low credibility of data and data analysis; inappropriate harvest regulations; low legitimacy of regulations; inadequate enforcement of regulations; overfishing; lack of attention to species interactions and habitat/ecosystem linkages; bycatch; habitat destruction; and failure to capture the full value of the resource. Four dimensions of partnership were discussed: their scope, scale, number of parties, and degrees of power-sharing. Partnerships were characterized in terms of their accountability, equity, and effectiveness, and conditions were identified which are predictors of successful partnerships which have these characteristics. Community development quotas were analyzed as a mechanism for partnering offshore and inshore fisheries, or of simply allocating fisheries access to communities which are otherwise disadvantaged in the marketplace. Umbrella regional organizations and regional boards were also discussed as ways of integrating offshore and inshore fisheries and promoting stewardship.

11. ACKNOWLEDGEMENTS

I am grateful to the Social Sciences and Humanities Research Council of Canada for supporting my research on co-management institutions over many years.

12. REFERENCES

Ames, E., S. Watson and J. Wilson. 2000. Rethinking Overfishing: Insights from Oral Histories with retired groundfishermen. In Neis, B. and L. Felt. *Finding our seas legs: linking fishing people and their knowledge with science and management*. Institute of Social and Economic Research, St. John's. p. 153-164

Berkes, F., George, P. and Preston, R.J. 1991. Co-Management: the Evolution in Theory and Practice of the Joint Management of Living Resources. *Alternatives*, **18**: 12-18

Clapp, R.L. 1998. The resource cycle in forestry and fishing. *The Canadian Geographer,* **42**(2): 129-44

Cordell, J. ed. 1989. *A Sea of Small Boats*. Cultural Survival Inc., Cambridge, MA. 410p.

deYoung, B., Peterman, R., Dobell, R., Pinkerton, E., Breton, Y., Charles, A., Fogarty, M., Munro, G., Taggart, C. 1999. Canadian Marine Fisheries in a Changing and Uncertain World. *Can.Spec. Publ. Fish. Aquat. Sci.*, **129**. 199pp.

Durrenberger, E.P. and King, T.D. eds. 2000. *State and Community in Fisheries Management: Power, Policy, and Practice*. Bergin & Garvey, Westport, CN. 250pp.

Dyer, C. and McGoodwin, J.R. 1994. *Folk Management in the World's Fisheries*. University Press of Colorado, Niwot, CO. 347pp.

Gendron, L., R. Camerand, and J. Archambault. 2000. Knowledge sharing between fishers and scientists: towards a better understanding of the status of lobster stocks in the Magdalen Islands (Quebec). In Neis, B. and L. Felt. *Finding our seas legs: linking fishing people and their knowledge with science and management*. Institute of Social and Economic Research, St John's. p. 56-71

Johannes, R.E. 1978. Traditional marine conservation methods in Oceania and their demise. *Ann. Rev. Ecol. Systems*, **9**: 349-364.

Johannes, R.E. 1981. *Words of the Lagoon*. University of California Press, Berkeley, CA. 245pp.

Klee, G.A. (ed.) 1980. *Word systems of traditional resource management*. John Wiley & Sons, New York. 290pp.

Langdon, S. 1999. Communities and Quotas: Alternatives in the North Pacific Fisheries. Presentation to the Pacific Marine States Fisheries Commission. Semiahmoo, Washington. 35pp.

Loucks, L., Wilson, J., Ginter, J., Fricke, P., and Day, A. 2002. Experiences with Fisheries Co-Management in North America. In D. Wilson, J.R.Nielson, and P. Degnbol, eds. *The Fisheries Co-Management Experience*. Institute for Fisheries Management, North Sea Center, Denmark.

Loucks, L, Charles, T. and Butler, M. eds. 1998. Managing Our Fisheries, Managing Ourselves. Gorsebrook Research Institute for Atlantic Canada Studies, Halifax, N.S.

National Research Council. 1999. *The community development quota program in Alaska*. National Academy Press, Washington D.C. 215pp.

Neis, B. and L. Felt. 2000. *Finding our seas legs: linking fishing people and their knowledge with science and management*. Institute of Social and Economic Research, St. John's 313p.

Pinkerton, E. and M. Weinstein. 1995. *Fisheries that work: sustainability through community-based management*. David Suzuki Foundation, Vancouver. 199pp.

Pinkerton, E. 1999. Factors in Overcoming Barriers to Implementing Co-Management in British Columbia Salmon Fisheries. *Cons. Ecol.* **3**(2) [online] URL: http://www.consecol.org/vol3/iss2/art2

Ruddle, K. and R.E. Johannes, ed. 1985. *The Traditional Knowledge and Management of Coastal Systems in Asia and the Pacific*. UNESCO, Jakarta Pusat, Indonesia. 313pp.

Schlager, E. and Ostrom, E. 1993. Property-rights Regimes and Coastal Fisheries: An Empirical Analysis. In T. L. Anderson and R.T. Simmons, eds. *The Political Economy of Customs and Culture: Informal Solutions to the Commons Problem*. Rowen & Littlefield Publishers, Lanham, MD. p 13-42.

Senge, P. 1990. *The Fifth Discipline. The Art and Practice of the Learning Organization*. Currency Doubleday, New York. 413pp.

Spoehr, A. ed. 1980. *Maritime Adaptations. Essays on Contemporary Fishing Communities*. U. of Pittsburgh Press, Pittsburgh. 161pp.

Stoffle, B.W. et al. 1994. Folk Management and Conservation Ethics among Small-Scale Fishers of Buen Hombre, Dominican Republic. In C. Dyer and J. McGoodwin. Ed. *Folk Management in the World's Fisheries*. University Press of Colorado, Niwot, CO. p. 115-138.

Thalenberg, E. 1998. Fisheries beyond the crisis. Canadian Broadcasting Company documentary film. 55 minutes. Box 500, Station A, Toronto, Ontario M5W 1E6

Wilson, J.A., Acheson, J., Metcalfe, M, and Kleban, P. 1994. Chaos, Complexity, and Community Management of Fisheries. *Marine Policy*, **19** (4): 291-305.

Yamamoto, T. and Short, K. eds. 1992. *International Perspectives on Fisheries Management, with special emphasis on community-based management systems developed in Japan*. National Federation of Fisheries Cooperative Associations and Japan International Fisheries Research Society, Tokyo. 527pp.

CHAPTER 8

FISHERY MONITORING, CONTROL AND SURVEILLANCE

by

Per Erik BERGH[1] and Sandy DAVIES[2]

[1] Ministry of Fisheries and Marine Resources, Namibia
[2] Marine Fisheries and Resources Sector Co-ordinating Unit, Southern African Development Community

1. INTRODUCTION 175
    1.1 What is monitoring, control and surveillance? 175
    1.2 A historical perspective 176
    1.3 The role of MCS in fishery management 177
2. THE MCS SOLUTION 180
    2.1 Strategy and plan 181
    2.2 Key strategic considerations 183
3. CORE COMPONENTS 191
    3.1 Before fishing 194
    3.2 While fishing 195
    3.3 During landing 198
    3.4 Post landing 198
4. FACILITATING FOR MCS 198
    4.1 Administrative options 198
    4.2 Information management and sharing 200
    4.3 Management system 200
5. ENSURING SYSTEM PERFORMANCE 201
    5.1 Assessing MCS performance 201
    5.2 Cost analysis 202
6. CONCLUSION 202
7. RECOMMENDED READING 203

1. INTRODUCTION

1.1 What is monitoring, control and surveillance?

In brief it could be said that monitoring, control and surveillance (MCS) is all about compliance to fishery management measures. This is of course a rather simplistic approach, but when the elements are analysed we see that they all lead towards this goal: monitoring gathers information on the fishery that is used to assist in developing and assessing appropriate management measures, while surveillance uses this information to ensure that these controls are complied with.

If a more precise meaning for MCS is required reference should be made to a definition developed by an FAO Expert Consultation in 1981 (FAO, 1981):

(i) **monitoring** – the continuous requirement for the measurement of fishing effort characteristics and resource yields;
(ii) **control** – the regulatory conditions under which the exploitation of the resource may be conducted; and
(iii) **surveillance** – the degree and types of observations required to maintain compliance with the regulatory controls imposed on fishing activities.

This definition may be helpful in clarifying the individual elements of MCS - but it is not a point to dwell. A definition alone will do little to assist a fishery manager grappling with the need to understand the role of MCS within fishery management or to assist them in finding a way forward towards an MCS solution. More important than a definition is the need to understand the core objective of MCS and to have some grasp of the options available to achieve this.

This objective of MCS is clear: to contribute towards good fishery management through ensuring that appropriate controls are set, monitored and complied with, (controls have been discussed in detail in earlier chapters of this Guidebook: technical measures (Chapters 2 and 3) and input and output controls (Chapter 4) are all considered as the 'control' element of MCS). This at the end of the day is what MCS is aiming for and whatever methods, tools, components or systems are used, the individual or joint outcome should contribute towards this objective.

On the other hand the options available for an MCS system and the various combinations of these options are almost limitless. They include a range of separate or interlinked components of hardware in varying degrees of sophistication, various levels and types of human resources (both linked and separate to the hardware), a whole host of approaches to implementation ranging from military type enforcement to community driven compliance programmes and then finally, once the system is developed, to even more choices of how to manage the MCS system and organisation. This chapter therefore aims to give an overview of the most common options available and an insight into some advantages and disadvantages associated with these different choices.

The use of the term MCS is sometimes criticised as being too wide and confusing in terms of concepts and functionalities in relation to the core function of the operations, compliance or law enforcement section of the fisheries management authority. This is essentially because the 'enforcement' section of the authority does not usually focus on the monitoring or control elements of MCS but rather on the surveillance and enforcement elements. However, for all the criticism the term MCS may receive it has become a common term used internationally and it offers a wider perspective that fits well with some of the more modern trends and approaches towards the issue of compliance and law enforcement. Therefore the use of the term MCS has been adopted for this chapter and all functions of MCS are considered, leaving the reader able to select the functional elements that relate to their needs or operational circumstances (i.e. who performs the tasks e.g. enforcement personnel, scientists or administrators is not specified, but the tasks are).

1.2 A historical perspective

Before we look further into the systems available today it is interesting to consider briefly the history of MCS and why for many fishery management authorities the MCS section of their organisation may be relatively new. In the early days of fishing some type of informal community or tribal management system often existed and this would usually include ensuring that fishers complied with certain accepted codes of behaviour. These informal codes of behaviour were based on community wisdom, philosophy and superstition, as to the best way to manage a fishery or water area that fell under the 'control' of one social group (community or tribe). If other social groups came into the given area, again informal codes of behaviour would dictate the way forward, or if these were not adequate, minor or major conflicts would break out. However, we see that the need for more formal and complex MCS systems is a relatively new concept that links very strongly to the United Nations Convention on the Law of the Sea and the establishment of the Exclusive Economic Zones (EEZs). Prior to this the majority of fishing activities within territorial seas could be viewed from the shore and this simplified MCS activities.

The MCS systems developed for the new EEZs were essentially developed as the implementing arm of fishery management, primarily to ensure that control measures, once agreed and adopted, were adequately implemented. Today this is still the core function of most MCS systems but due to the integrated approach to fishery management, encouraged through many international instruments and specifically the FAO Code of Conduct for Responsible Fisheries, a far greater and more linked role for MCS is emerging. In this new role MCS strategies now include the need to contribute towards the development of management plans (and therefore control measures) through the provision of information that is key to the evaluation of different management measures. MCS systems are also becoming active in the promotion of compliance by fishers through user participation, rather than following the old focus on the enforcement of controls. These two new trends are changing the approach of MCS in many parts of the world and are bringing it closer to other sections of fishery management and also to the fishing communities.

1.3 The role of MCS in fishery management

Often the concerns of MCS have been overlooked in the development of management strategies and plans in light of the belief that good fishery management is considered synonymous with good science. That is to say that as long as suitable scientific analysis and modelling was backing the choice of management priorities and measures then the need to successfully implement these measures (to obtain a high level of compliance by the fishers) was ignored. However, as a result of many unsuccessful management regimes that were based primarily on scientific assessment, the need for a more balanced approach is becoming increasingly evident and popular: an approach that considers compliance with conservation-based measures as essential for proper management of fishery resources. The emerging picture in modern fisheries management is therefore one of interlinked and compatible systems that provide feedbacks and checks to the management strategy – MCS is one of these systems.

Modern fisheries management is therefore placing MCS strategy, planning and activities at a far more central and integrated place around the table of fisheries management (see Figure 1, Chapter 1). For example, in Canada enforcement officials now regularly attend consultative meetings with industry and actively participate in the development of management plans. It is still clear that when the objectives of the fishery are chosen (biological, ecological, economic or social) concerns over MCS will rarely apply, as the objectives are related to the direction given by the national and fisheries policy. However, when we move down a level to the discussion on the alternative management strategies (including the selection of the management measures) that will be adopted to implement these objectives, there are various aspects related to MCS that should be considered. Below, some points are listed that indicate the type of questions that the MCS representatives should be asking to ensure that the MCS concerns are considered when accessing any proposed plans:

(i) what are the practical requirements needed to implement the management measures (this should be considered from the monitoring, surveillance, compliance and enforcement points of view) and are these available;

(ii) an evaluation of any previous records of success or failure of management measures should be made (preferably quantitatively but even in a qualitative manner if no data are available) and the results considered in light of any proposals;

(iii) what are the factors that will encourage compliance rather than demanding enforcement and what are the requirements to develop these – are they feasible;

(iv) the consequences of non-compliance (i.e. violations of the set controls) must be considered in relation to the effect that these will have on the status and viability of the fishery, therefore the level of compliance that is required in order to support the management plan should be considered;

(v) what is the cost of these management measures and/or non-compliance from both a financial and resource perspective and, from a financial perspective, who should cover these costs, government, industry, or both?

Potentially, illegal fishing or illegal fishing activities could compromise the implementation of management plans and can, in extreme cases, undermine the rational exploitation of the resource. For this reason the FAO International Plan of Action to Prevent, Deter and Eliminate Illegal, Unreported and Unregulated Fishing[1] was developed and adopted by COFI in March 2001 (Chapter 1). Also a management plan (however simple) that cannot be properly implemented may damage the credibility of the fishery management authority and be detrimental to the management of other fishery resources. It is therefore important to try to ensure that management plans can be properly implemented and that non-compliance is kept to an acceptable level.

For example, the use of total allowable catches (TACs) as a means to control catch levels implies that all landings must be monitored and catch by species recorded in close to real time (e.g. through logbooks and sampling or a complete landings monitoring programme). Also adequate steps are required to prevent discarding at sea of the target species and the unregistered transhipment of catches. It must therefore be asked; can the MCS organisation implement these required checks; or can the organisation be realistically developed to do so? Another example could be in relation to an effort control of the number and fishing capacity of vessels. Effort controls are generally considered less expensive to implement than output controls but require accurate fleet registration, close monitoring of fleet performance and of technical or operational developments that may affect efficiency (see Chapter 4, Section 6). Again, the question must be asked can the MCS organisation do this? Even if we consider one of the simplest control measures such as closed seasons or closed areas, these require the ability to monitor the closed times and areas (e.g. through vessel or plane patrols) or to develop voluntary compliance in such a way as to ensure that the management measures are adequately implemented, again it must be asked can this be done?

---

[1] The details of this IPOA can be found at http://www.fao.org/fi/ipa/ipae.asp

As well as considering the requirements that fishery management plans have of MCS, what MCS requires from management plans should also be considered. MCS activities must relate to specific management objectives, therefore clear management statements are required to develop MCS systems to the appropriate levels and at an appropriate cost. In addition to management objectives, information about management priorities, management measures and the available resources will also be needed. Even if full management plans are not in place an indication on these issues is required as the MCS strategy will aim to marry the priorities for the fishery with a practical approach to ensure an acceptable level of compliance, and also to balance the reality of limited enforcement resources with the expectations of industry and, to some extent, the other branches of the management authority.

Figure 1 gives a simplified diagrammatic representation of the main information links between management, science and the MCS functions of a fishery management authority. It does not aim to give the complete picture of fishery management but to highlight the main interactions and feedbacks. It is important to note that the three functions depicted (management, science and MCS) are not necessarily synonymous with sections of the same name within the fisheries management authorities, (e.g. monitoring can be performed by either the scientists or the enforcement officers).

An example of the links and feedbacks in the diagram could be that the link joining the monitoring to scientific research could relate to the fact that the scientific section analyses the influence of the management measures on the fish stock and the fishery, while both the MCS and scientific sections provide information for the analysis, and the MCS section provides details on how compliant the fishers are to this management measure. For example if a mesh restriction of 150mm is in force on a fishery, the scientific section will evaluate the influence this has on the catch composition (possibly using information collected by the MCS section). This information is then extrapolated through modelling into predictions of fish size (and age) for the entire catch of that species, but of course these predictions are assuming that there is 100% compliance to the 150mm mesh restriction. It is therefore the task of the MCS organisation to ensure that the management measures are complied with or (and more realistically) to inform the scientists of the estimated level of non-compliance. With this information the scientists are able to adjust their models to reflect a more accurate estimate of the size structure of the fish caught. This information will then be passed to management in two forms; firstly in the link between scientific research and fisheries management as scientific predictions on the status of the stock and as advice for future restrictions or management measures; and secondly through the link between MCS (strategy) and fishery management as information on non-compliance. If non-compliance is high (that is the controls are regularly being violated) it is an indication to management that the controls are unsuccessful.

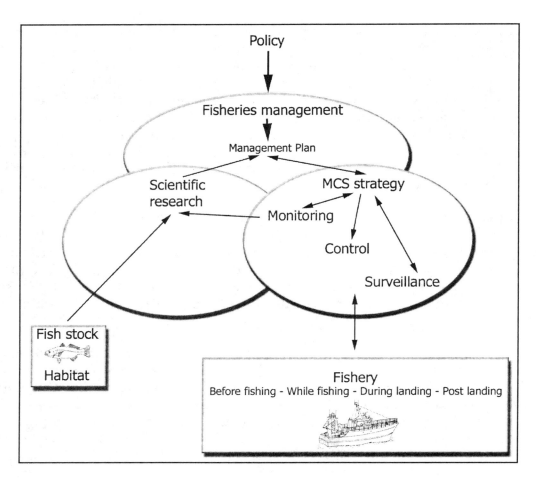

Figure 1 The main links between MCS and fishery management

2. THE MCS SOLUTION

We have seen that MCS is an integral part of fisheries management, both requiring information to set the strategy and plan and also feeding information into the management system to assist in producing management decisions: or to put it another way the type of controls set will influence the monitoring and surveillance, while the monitoring and surveillance should influence the types of controls set. It is now necessary to consider which factors play a role in the design of an MCS solution for a given fishery.

First and foremost there is no unique MCS solution for all fishery situations, nor are there inherently right or wrong approaches to the implementation of MCS systems. MCS systems should be developed for either specific fisheries or a group of interacting fisheries (in terms of ecological, fleet or management interactions). The MCS system chosen will be shaped by a variety of factors and the main ones are discussed in this section. The factors include aspects important to all three elements of monitoring, control and surveillance. Different factors will hold varying levels of importance depending on the situation, for example a multi-species artisanal fishery will have different priorities to a large-scale single-species industrial fishery. However, it is worth considering all of the points made in this section in order to evaluate their importance.

2.1 Strategy and plan

In designing a MCS strategy key strategic considerations are required and these are discussed in section 2.2. The following points although not key considerations may assist a manager in developing the strategy.

(i) An MCS strategy is vital if any overall cost benefit is to take place within the fishery. It is often difficult to stop an activity or to start one within an established organisation and for this reason analysis of the economic inputs and outputs from a fishery are required to determine what activities to plan for, what to stop, what to upsize and what to downsize.

(ii) One of the struggles that MCS managers face is how to balance the need for a flexible strategy and plan that is able to react to the dynamic nature of fish, fishers and fisheries and how to manage this flexibility within a system of annual planning for financial, human and hardware resources. One solution in more traditional MCS systems (e.g. vessels, aircraft and observers) is to identify which activities may need to be most flexible and always to plan these with room for adjustments and changes and if possible to include a mid-year review of the plan. Another solution is to adopt a MCS system utilising more modern components such as a Vessel Monitoring System (VMS) or remote sensing as these are by nature more flexible.

(iii) A breakdown of the activities into two core areas; firstly a component covering enforcement including policing and deterrence; and secondly the monitoring and compliance component including stakeholder consultations and sector awareness actions, can be a useful approach to dividing activities.

(iv) Usually economic return from a fishery is one of the objectives, therefore it is interesting to consider, for example, that achieving 100% compliance may cost more than the economic returns from a fishery, whereas 70% compliance could be both affordable and the remaining 30% non-compliance could be planned for in the management plan and therefore not become a threat to the sustainable use of the resources. The point of interest is that when developing a strategy for monitoring or surveillance a balanced and realistic approach must be considered. It should also be noted that the level of compliance aimed for is a strategic decision required separately for each fishery.

(v) Both short-term and long-term strategies are important for good fisheries management and it is vital that policy, strategies and plans have the same objectives to work towards.

(vi) A well-designed strategy will include the consideration of assessing the performance of the MCS system against the targets (refer to section 5).

(vii) Most of the larger fishing nations rely on some level of consultation with resource users; however the level of participation that is allowed and encouraged is a key question when developing the MCS strategy. It is worthwhile to consider some of the arguments for greater participation or sharing of power and the long-term value of compliance over enforcement as discussed in 2.2.7 of this Chapter and in Chapter 7, and also the current international recommendations that are encouraging this type of participation such as Paragraphs 6.13, 7.1.2 and 10.4.1 of the Code of Conduct. The process should also include groups other than fishing interested parties, such as environmental groups. Long-term strategies should aim to resolve conflicts between different interest groups (e.g. artisanal and industrial fishers, trade or environmental organisations).

(viii) Finally when revising or developing a strategy the following simple questions should always be asked.

- What is required in terms of the fishery you are managing?
- What is feasible in terms of the legal framework?
- What is realistic in terms of available resources?
- What is practically possible to implement taking into consideration the political situation and the interested parties involved in the fishery?

An example of a simple strategy and plan for the MCS system could be if we imagined the management plan for an extensive artisanal fishery with very limited information on catch or effort, but the knowledge that decreasing catches and heavy exploitation in certain areas had lead to the introduction of a new minimum mesh regulation for a certain gear, a ban on beach seines and two designated no-fish areas. The country has a national employment policy and limited financial resources available to manage the fishery. The fishery is not of high economic value but important for local employment and food security. This information is the type of information that can come from a management plan to the MCS organisation, perhaps in more detail or perhaps not.

A possible strategy for implementing these measures would go along the lines that in the short term emphasis would be placed on two aspects, that of data collection and that of encouraging voluntary compliance through community self-monitoring and fisher obligation (developed through an awareness campaign). The data collection should employ local staff on a part time basis supported by a team of supervisors. These supervisors would also spearhead the awareness campaign and the community self-monitoring programme. The long-term strategy would look into possibilities to create alternative employment through the tourist industry for fishers. The strategic goal would be that by year three, quality catch and effort data would be available to allow for simple stock assessment, by year three compliance by the fishers for gear and area restrictions was over 80% and by year 5 alternative livelihoods have been identified for 5% of the fishers.

The plan would then give operational details to the strategy. For example in month 1, 20 data collectors are to be employed from various points in the fishery. A training course of one week is to be given to the data collectors with instruction given by a team of 3 inspectors from the MCS organisation that will form the support to the data collectors and also plan and implement the awareness campaign. A frame survey will be carried out in month 2 using the data collectors and supervisors and initial introductions and information dissemination to the community will occur. In month 3 a sampling programme for data collectors will be designed in light of the frame survey data and further training given on sampling techniques and form completion etc. By month 5 the data collectors will be sampling one day per week at their allocated beaches. The supervisory team will visit data collectors on a regular basis to collect forms and to discuss the work. The awareness programme will start in month 4 with community meetings and radio transmission on the need for conservation in fisheries .... and so on until an overall MCS plan is developed for the year.

The example above, although very simplistic and limited, shows how the three levels (management plans, MCS strategy and MCS plans) are linked. It is important to have both a strategy and a plan within the MCS solution as they assist in many ways in ensuring a sense of objective and purpose to often very isolated and separated activities (for example due to the strategy it will be relatively easy to explain to data collectors where their role fits into the overall plan for fisheries management, without it this link is less tangible). A MCS strategy is therefore necessary to give the organisation clear indications in relation to priorities, resource allocation and human resource development, while the plan turns this into practical reality. It can take years to build and train an organisation to a level of satisfactory performance, a point that underlines the need for long-term strategies and detailed planning.

2.2 Key strategic considerations

Following on from the last section, eight key strategic considerations are now discussed. These considerations are important in the shaping of the MCS strategy and some aspects may be relevant for the MCS plan.

2.2.1 Type of fishery

Industrial or artisanal

Industrial fisheries require integrated cost-effective MCS solutions – these will usually include various components such as vessel registers, observer programmes, VMS and patrol vessels and aircraft. Often the emphasis is on enforcement rather than compliance although this is changing and is discussed further in section 2.2.8. The monitoring aspects of industrial fisheries are generally easier than in artisanal fisheries as vessel logbooks can often be implemented, VMS and observers can be placed on larger vessels and the landing of fish is usually through certain ports that facilitate landings monitoring.

If foreign fleets are involved in the fishery it is generally important to maintain a good MCS system with an emphasis on deterrence and enforcement but also ensuring adequate monitoring of catches. Voluntary compliance by foreign vessels is more difficult to achieve even when long-term commitments to the fishery exist. National fleets will generally be more amenable to voluntary compliance and this can be developed through participation in the fishery management process (see Chapter 7).

In artisanal and small-scale fisheries the combination of large numbers of fishers, mixed gears, migrant fishers and the proliferation of landing points makes MCS a very complex task. Often the most appropriate approach to MCS in these fisheries is through the community-based approach. However, alternatives such as government data collectors sampling at landing sites, combined with frame surveys and possibly some enforcement presence can also offer low cost solutions.

Multi-user fishery

When more than one type of fisher is targeting a fishery (e.g. artisanal, industrial or recreational) often the MCS effort will target the user that offers the biggest threat to the fishery in terms of financial loss or biological damage. It is more often than not the case that the larger the vessels the greater the potential crime, but if a large amount of small vessels or gears are violating controls the cumulative effect can also be significant. It is important to consider all the users, especially to ensure that the monitoring programme covers the whole fishery, perhaps through a sampling programme.

An example of a multi-user fishery is the Namibian coastal line fishery. It is a multi-species fishery and is exploited by commercial linefish vessels, day-trip skiboats (commercial and recreational), recreational anglers and subsistence anglers. The majority of MCS effort is targeted at the commercial linefish vessels as they are catching the largest volume of fish and they potentially offer the largest biological threat to the fishery. They are controlled through a system of licences and closed areas and monitored through logbooks and monitoring of landings by inspectors. However, effort is also given to monitoring the other resource users where the control measure is a bag limit that is applied for the recreational and subsistence sectors. These are monitored at varying intensities during the year mostly corresponding to the tourist season, by a combination of awareness campaigning, beach inspections and road blocks.

Offshore fisheries may also require a strategy for interaction between fishers (e.g. in a demersal longline and trawl fishery) or between fishers and other users (e.g. oil prospecting or drilling operations), again participation by all users is a useful approach to this type of MCS issue.

Gears

As discussed in Chapter 2 of this Guidebook, the managers' toolbox includes both passive and active gears and the type of gears used in the fishery will influence the type of MCS required. In general passive gears are easier to monitor and control than active gears that require more mobility, equipment and more complex detection systems. When an effective MCS system is in place potential poachers are far more likely to use active gears than to risk the potential of being trapped when returning to collect passive gears. In Norway, the police or coastguard often lie in wait for returning salmon fishers that have left illegal nets (i.e. unlabeled and missing the licence number details) in the water. When offenders are identified they receive heavy fines and the nets are confiscated.

Multi- and single-species fisheries

Multi- or single-species fisheries will also have different demands on the MCS system. Single-species fisheries will often have more complex control measures to implement but on a more uniform fishery that will make monitoring more simple. Multi-species fisheries may have less complex controls but a large variety of fishing methods and vessels that require considerable effort to monitor adequately, in order to gain accurate information on catch and effort.

2.2.2 Type of management measures

Use rights

There are many different management strategies that combine different management measures, and linking these management measures to a feasible MCS strategy is important. As a starting point it is very difficult to ensure compliance to a specific management measure in an open access fishery and this is a reason to encourage the implementation of rights-based fishery management strategies. The type of access rights that can be used (e.g. territorial rights (TURFs) or limited entry access rights) are discussed in detail in Chapter 6 of this Guidebook and these will be important in the type of MCS system developed. Ensuring that only the allocated fishers are exploiting the resource and that no poaching is taking place is a core MCS activity that is best addressed in the strategy. Depending on the value of the fishery and the threat by poachers it may be necessary to put major effort into surveillance activities. In the south west Atlantic, the squid fishery is a seasonal fishery and the fishing fleets follow the squid through its migration across highseas, and through two EEZs. Within the EEZs poaching can be high especially when the squid is located around the boundary area. The Falkland Islands Government responds to this by putting patrol vessels and aerial surveillance in the EEZ boundary area throughout the period of squid migrations and aims to intercept and arrest poachers.

Input, output and technical controls

Input controls relate to effort control and this can be broken down into two types of controls related to the number and size of vessels (capacity control) and the time spent fishing (usage control). Output controls relate to limiting what is caught through limits of TAC, bag size or limiting bycatch (refer to Chapter 4 for a fuller description). Broadly speaking effort control is easier to enforce. Vessels and fishers need to be monitored for compliance to the effort control (such as numbers of lines, areas fished or vessel horse power) but usually fishers are more willing to provide catch and effort data as there is no benefit from giving falsified information.

Closed areas and seasons (Chapter 3) are a means for controlling effort and also for limiting the possibility of certain bycatch, size or spawning condition by means of limiting the area available for fishing. This type of zoning lends well to self-policing or community involvement in self-surveillance. If a fishery is seriously in danger or recovering from overexploitation, such as some of the reef fisheries around the islands of the southern Indian Ocean (recovering from disruptive exploitation methods such as poison and dynamiting) then it may be appropriate to close an area completely. Closed areas are generally easier to enforce than gear or catch restrictions or 'no take' areas.

Output controls on the other hand are the hardest to control as they require accurate figures on catches (usually by species) that require either complete monitoring or a detailed inspection programme to support logbook, landing or processing data. If a high level of potential violations is expected, a strong deterrent to fishers will be needed combined with an evaluation of the level of non-compliance.

2.2.3 The legal framework

The domestic legal arrangements within a given country set the framework and basis for the operational activities of the MCS system (see Section 9, Chapter 1). It is therefore important that both those developing and those operating an MCS system understand these legal arrangements and the mechanisms to change them. From an operational view point the conduct of fisheries investigations and the preparation for and conduct of prosecutions are two additional areas relevant to those operating an MCS system.

At the top of the legislation is the constitution of each country, followed by Acts (or similar instruments). There will most likely be an Act relating to Fisheries and one that determines the Maritime Zones. These are both of importance to the MCS strategy. The Act relating to fisheries will be the most important and it should provide details on definitions, management regimes (including the designation of power to the appropriate government authorities to determine national MCS policy), powers of MCS officers (inspectors, observers etc) and details on the offences that can be made under the Act. Following the appropriate Acts there will most commonly be some form of regulations or system for providing lower level supporting legislation to the fishery, this may include licences and other legally binding documents where the rules and regulations of the individual fisheries are set out.

In order to design an MCS strategy that will be possible to implement, it is important to consider this legal basis and to aim to utilise the legal strengths of the system by selecting a strategy and MCS components that will be able to work effectively within the given legal framework. The actors that may need to be considered in legal considerations include the state actors, international organisations and non-state groups (e.g. industry associations, NGO's, vessel masters, owners and fishers). The key facts that the legal framework will establish are who can fish, where, what, how much, with what, and where it can be landed. Understanding how these work and how the law is set up to deal with actions that do not legally agree with what is set out on these issues is something that every fishery manager must know.

2.2.4 Human resources

People are the centre of any MCS organisation or operation. No technology, strategy or plan will be able to replace the demand for quality personnel. Therefore consideration of the human resource requirements of an MCS solution is required. At the strategy level a realistic evaluation of the personnel available to the organisation is required in relation to the MCS needs, the financial resources, the time available and the feasibility for long and short-term training. The following points could be considered.

(i) Knowledge levels: what are the minimum knowledge levels required for the different personnel tasks or professions?

(ii) Recruitment procedures: what are the criteria used for recruitment and will these be suitable for the MCS needs?

(iii) Probability of corruption: what is the potential for corruption among MCS personnel and are there any anti-corruption initiatives that could be implemented?

(iv) Training capability and capacity: what level of training can the organisation provide for personnel and what external training will be required, how long will it take?

(v) Political, social and policy requirements: is your organisation aiming for a labour intensive industry or is high technology and efficiency more important?

Knowledge about the dynamics of the fisheries, possible infringements and effective use of MCS resources are the main keys to success in a MCS operation and if this is lacking the strategy must address how to get it. Basic knowledge will be needed immediately if the organisation wants to gain respect from interested parties and donor technical support may be one option to initiate a professional and functioning operation if such knowledge is not available. In order to assure the long-term sustainability of the organisation training must form part of the strategy and plan. This training should ideally be a well-designed training plan for all levels of staff, which runs throughout their career structure.

When considering the organisation's approach to a human resource strategy the following may be useful.

(i) It is worthwhile considering if a well-trained, better-paid and smaller work force would result in higher productivity and a more effective organisation than a less competent larger workforce.

(ii) Training should ideally be officially acknowledged, for example through permanent employment, higher rank, bonuses or higher salary. This is important for motivation, sustainability and recruitment in the organisation and should be reflected in a human resource development plan.

(iii) Practise has shown that the most effective way to train for lower level jobs is through vocational modular training preferably based on adult learning principles. These training programmes (for example for observers, inspectors, clerks and radio operators) can often be taught by more senior personnel in the organisation and developed for specific local situations.

(iv) In relation to in-house training courses it is also important that quality criteria are demanded from instructors to ensure that a certain level of quality is maintained in the teaching.

2.2.5 Financial requirements

Cost effectiveness

Cost effectiveness is a primary consideration for all MCS systems and a comparison between the costs and benefits from different MCS options is required. The MCS strategy will need to provide clear guidelines on the financial resources available and on the approach to allocating these. Generally, if the costs of the MCS solution exceed the expected financial and other benefits of the MCS interventions, then alternative, less costly options should be explored. However, there are always exceptions to the rule such as when a country values a resource beyond its immediate financial return for social or historical reasons. To aim for a cost effective solution may appear to be an obvious conclusion, but it's surprising how often this is overlooked. In the 1990's the United States spent approximately $80 million on the surveillance of foreign fishing operations while collecting only $41.5 million annually from the same fishing fleet. In contrast in the small coastal state of Costa Rica, the cost of a modest enforcement programme for the tuna fishery was calculated to be about 50% of the expected revenue from this fishery. Another example is Namibia which collected N$120 million (US$15 million) from the fishing industry in 1999 while the cost of the MCS organisation was estimated to be N$66 million (US$8 million). This indicates a sound and sustainable organisation well proportioned to the financial income of the sector.

Who should pay?

The questions that must be considered at the strategy development stage (if this has not happened previously) is how can enough income be generated to meet the needs, who should pay and how should they pay? There is an increasing trend to recover costs from those active in the fishery, such as the fishers, the boat owners, the port owners and the fish processors: by determining which interest groups benefit from the fishery, costs can then be appropriately attributed and recovered. As fishers are usually the primary beneficiaries of MCS programmes it is worthwhile considering how much of the cost they should bear. It is recommended that this share may best be increased incrementally with time – this has the dual effect of encouraging more compliance because increased compliance implies reduced costs and also encourages the industry to internalise the costs of their sector. As an alternative, it may be feasible to tailor the MCS programme around fishers' ability or willingness to cover costs by linking specific management measures to MCS programmes. For example, the "cost" of an individual quota system (as it may require additional monitoring resources) would be higher than that of a competitive fishery and as the fishers allocated quotas would benefit from a well implemented system, they should also cover some or all of the additional costs associated with the quota system.

Donor support

Special provision is given to developing countries in the Code of Conduct (Paragraph 5.2) where countries, relevant international organisations, whether governmental or non-governmental, and financial institutions are called on to assist developing countries in areas including financial and technical assistance, technology transfer, training and scientific cooperation. Funding for fisheries management in many developing countries relies heavily on donor assistance and this assistance is in many aspects the only solution for any organisation lacking resources and expertise. There are some pitfalls related to this type of support: common examples include the many developing countries that have fallen victims to eager donors that haven't considered the receiving countries capacity to operate and maintain their generous gifts of expensive hardware. Often equipment such as patrol vessels or planes cannot be utilised due to lack of resources and end up tied up at jetties or in hangers. In such cases, greater benefit would have resulted from a more moderate MCS-system with consideration of long-term costs. It is therefore important to ascertain the long-term commitment of the receiving government before accepting any technical assistance in the form of hardware.

Low cost options

When considering costs it is worthwhile asking the question do lower cost options exist? Normally the answer is yes they do. Larger commercial fishing operations including domestic and/or foreign vessels do not necessarily require patrol vessels and planes as part of the MCS system. A VMS combined with a certain degree of observer coverage can also do the job. The main cost of the VMS system (vessel unit) can be borne by the industry, while a simple observer compliance and data-collection programme can be established to compensate for the weaknesses of VMS. Vessels participating in the fishery can be channelled to certain harbours or checkpoints before leaving the fishing zones for control purposes by inspectors. This type of system is able to address a wide combination of management measures on an already licensed fleet. In order to improve the level of compliance, the above system can be combined with a low cost awareness and participation programme to encourage fishers to be engaged in the decision-making for the fishery. Additionally or alternatively, if unlicensed vessels are a serious problem, small private aeroplanes carrying one fisheries inspector could be leased (for example twice a week) and this in combination with low-cost patrol vessels would have a deterrent impact on illegal activities. Alternatively assistance from the Navy or Coastguard may be a possibility.

Regional and bilateral strategies

Another option for cost saving is to incorporate an MCS strategy and operations plan into bilateral or regional fisheries agreements. This approach is encouraged in Paragraph 7.7.3 of the Code of Conduct. A successful example of this is a low cost solution called 'no force' that was developed by the South Pacific Forum Fisheries Agency and implemented in 1986. The concept is built on the principle of voluntary catch reporting from the fishing vessels, regional sharing of enforcement costs, regional sharing of catch and compliance information and use of "good standing" for granting of fishing rights. This system has joined together 23 countries and territories covering 30 million km² of ocean and one of the world's most productive tuna fisheries, through a regional MCS strategy and plan. Bilateral or regional co-operation apart from the potential cost saving angle can also be of great value when fishers are migratory either due to trans-boundary or migratory fish stocks or simply due to their searching strategies to locate fish stocks to exploit. There are many further examples of regional co-operation (e.g. the Indian Ocean Tuna Commission, the West African Sub-Regional Fisheries Commission, the Organization of Eastern Caribbean States Fisheries Unit). Some are inevitably more successful than others, but in recent years more success stories are emerging from this type of shared management.

2.2.6 MCS dimensions

In considering the area and dimensions that MCS covers, it should first be noted that MCS is related to the fishery (this includes the fishers and fishing related activities) and not to the fish stock *per se*: fisheries are managed by managing the fishers not the fish. So MCS relates to routine fishery operations, this includes four key dimensions; before fishing, during fishing, landing the fish and post landings (Figure 1).

These four dimensions should be considered when designing an MCS strategy and plan in order to obtain the optimal level of monitoring and surveillance at the least cost. For example if all the MCS effort is placed on the 'during fishing' dimension this would not facilitate any crosschecking or validation across dimensions. Ideally the aim should be to spread the monitoring and surveillance across the four dimensions. For example in the European Community cross-checking is made through logbooks, catch and effort reports, VMS, patrol vessels and planes during fishing, landing declarations at point of landing and sales notes at post landings.

There will inevitably be trade-offs between different combinations of solutions such as the choices between covert or overt surveillance, education of interest groups or traditional enforcement, total monitoring or sampling. For example, a regular presence by MCS platforms such as patrols vessels will act as a deterrent to discourage potential violators from carrying out illegal operations, but equally it is important that an initial inspection performed in an area arrives as a surprise to the crew of the fishing vessels. These two strategies are obviously at conflict and it is therefore important to find a balance that suits the objectives of the operation.

2.2.7 Targeted MCS

The use of targeted MCS is an important strategic consideration that can have a large impact on the cost effectiveness and the efficiency of an organisation. When an individual fishery covers a large physical area or the total number of fisheries managed by one-authority covers a large area, the MCS resources of that organisation will often be over stretched. Random checks as part of a sampling strategy may be sufficient to get the required data for monitoring, but often the surveillance effort that is targeting enforcement must be aimed at known or suspected offenders, that is, it must be intelligence driven. This targeting of routine offenders rather than the occasional or opportunistic offender is important to catch offenders and also as visible deterrent to potential offenders.

This intelligence driven enforcement is also known as adaptive operations. The information usually comes from the fishing community itself and is therefore part of a co-operative or participatory management approach. Encouraging reporting can be a difficult task: fishers will often feel loyal to their colleagues (one day their safety may depend on them). Therefore to encourage fishers to report, the following ideas may help; a code of ethics for fishers, education to the fishers about their role in the management system, easy reporting methods (e.g. Western Australia's 24 hour telephone line or the internet reporting system for the Southern Ocean toothfish), a reward for the information, clear administrative systems and legislation and, perhaps most importantly, the nurturing of good relationships between the inspection officers and the fishers.

2.2.8 Compliance or enforcement

Many fishers operate in an environment rigorously controlled by the authority. The area they work in is however often isolated without witnesses or law enforcement units present. Fishers are therefore easily and frequently tempted to violate the regulations designed, as they often see it, specifically to restrict their effectiveness. In addition, even in the most advanced and complete enforcement systems, fishery inspectors can rarely be everywhere due essentially to cost limitations. So what can the fishery manager do? The answer given more and more often these days is to balance the enforcement and compliance aspects of the MCS system, to encourage an environment where maximum compliance from fishers occurs and to use enforcement in areas where voluntary compliance is not successful or requires support.

The balance between compliance and enforcement is a question that must be considered at the strategy stage. It is not something that is only applicable for artisanal or small-scale fisheries: voluntary compliance has a role to play in all MCS strategies and it is generally considered to be the positive output of adopting a participatory approach, which is the essence behind Paragraph 7.1.2 of the Code of Conduct.

Legitimacy

The assumption with legitimacy is that people are more inclined to obey rules that they feel are legitimate (rightful, justifiable and reasonable). Therefore, there needs to be a perception of fairness in legislation if it is to become effective. Creating a sense of legitimacy towards the management strategy or any particular controls will depend on many factors, such as:

- the content of the regulation itself – how does it compare with the view of the fishers;
- the distribution of the regulations – are they equitable;
- are other partners (fishmongers, processors, recreational sector etc) carrying a fair burden of the enforcement;
- were the fishers involved in the formulation of the controls and regulations;
- is the implementation transparent;
- do the fishers feel an ownership towards the management;
- is there a good dialogue between the authorities and fishers?

One way to ensure these is through a balanced strategy that is open for all to see.

Legitimacy does not just apply to the legislation but also to the perception of the fishery management authority. If the public perception of the authority is low in terms of technical skills, corruption, laziness and public arrogance, this will have an effect on the overall compliance by fishers – as well as being internally very destructive!

Deterrence

Deterrence is another way to increase voluntary compliance and it mainly relates to the severity and certainty of sanctions. Illegal activities must be unprofitable, and more importantly it is vital that fishers can not get caught for a violation and still gain from the crime. This may appear obvious but there are many fishery management regimes that have not addressed this issue adequately.

If voluntary compliance is an objective then crime mustn't pay (Code of Conduct, Paragraph 7.7.2): if the deterrence is high enough then compliance is encouraged. In Western Australia the rock lobster fishery is a high value fishery where industrial fishers are given 'black marks' for serious offences. If three black marks are given in a 10-year period then their license is cancelled. This high penalty system has ensured high compliance.

Participatory management

Where does the obligation lie to ensure that the management measures will be complied with? In answer to this question and due to the failure of many traditional enforcement driven MCS systems, participatory or co-operative management methods are becoming more popular as a means of fishery management in partnership with other interested parties (including of course the fishers). There are many advantages in involving willing fishers; their understanding and knowledge of the fishery will increase, the chance of violations due to lack of knowledge will decrease and hopefully their desire to comply and assist in ensuring that others comply with controls will increase.

Community management is a term that generally refers to the involvement of small-scale or artisanal fishers in the fishery through the community structure, while participatory management refers to all types of fisheries and includes community management. Community management does have a special place to play in MCS and for many countries it is the most feasible option to encourage compliance. For example even high penalties and deterrence will not be effective if fishers are either financially desperate or hungry – in these cases the number of violations will increase. This may be a serious and difficult situation for a fishery manager and often only community intervention will be able to influence fishers. Chapter 7 of this Guidebook gives a fuller discussion of the subject.

Apart from the obvious advantages of voluntary compliance from a biological point of view, it also has significant financial implications for the MCS organisation: if the compliance is greater then costs of enforcement are less. However, it is important to note that there are cases where enforcement is essential and certainly voluntary compliance is not the best route to follow in all cases. On the negative side, voluntary compliance tends to take longer to implement and for the results to become apparent – this may spell disaster if violations are critical to the sustainability of the stock where the best option may be immediate enforcement action.

3. CORE COMPONENTS

This section considers the possible core components of an MCS system with a focus on physical components and hardware. Information is included on the objectives of each component and the ability it has to implement different control measures (Table 1). Selection of components will relate to the MCS strategy including cost considerations and the points made in Section 2 of this Chapter and Table 2.

New technology may offer the possibility of improved MCS systems and enhanced cost efficiency, but it should be noted that technically advanced MCS components may take years to develop into efficient instruments. The level of compliance required, knowledge, MCS experience and running costs should be given serious considerations in the planning phase before implementation. If new, more technological solutions are chosen it is important that old procedures and working practises are revised to take advantage of the new components. Another challenge is to manage organisational changes: changing old routines and analysing ways to improve effectiveness and efficiency are the only way to fully utilise the potential of new developments.

Table 1 Comparison of the effectiveness of different MCS components to implement control measures

Dimension	Component	Effectiveness of element for management controls			Detection of unlicensed vessels/fishers	Power of arrest	Cost
		Input	Output	Technical			
Before Fishing	Clearance / issue of documentation	Medium	None	None	No	Yes	Low
	Vessel clearance	Medium	None	Low	No	Yes	Low
While Fishing	Logbooks	Medium	Medium	Low	No	No	Low
	Patrol vessels	Medium	Medium	Medium	Medium	Yes	High
	Patrol planes	None	None	High	High	No	Medium
	Helicopters	None	None	High	High	Yes	High
	Observers	High	High	Medium	Low	No	Low/ Medium
	VMS	Medium	None	High	No	No	Low/ Medium
	Satellite Imagery	None	None	Medium	Medium	No	Low/ Medium
	Beach patrols[2]	High	High	High	High	Yes	Low
	Navy or coastguard	Low	Low	Low	High	Yes	High
During landing	Catch monitoring	None	High	None	No	Yes	Low
	Transhipment monitoring	None	High	None	No	Yes	Low
Post landing	Market and sales monitoring	None	Medium	None	No	Yes	Low
	Export monitoring	None	Medium	None	No	Yes	Low
	Roadblocks and transport monitoring	None	Low	None	No	Yes	Low

---

[2] Only beach related fishing activities

Table 2. Advantages and disadvantages of different components of MCS

Component	Advantages	Disadvantages
Clearance / Issue of documentation	Ensures valid documentation among the fishers and provides opportunity for briefing of captains.	Can only be performed on vessels calling at national ports with an MCS presence.
Vessel clearance	Good source for information about the fishery. Controls in relation to e.g. engine size, fishing gear can be conducted.	Fishing gear and other equipment may be hidden.
Logbooks	Can be used onboard any fishing vessel in any language. Keeps historical track on catches and positions. Cheap	Poor literacy rate by fishers may be an obstacle in certain fisheries. Quality of data will depend on fishers' motivation.
Patrol vessels	Provides at-sea verification of fishing gear, discards, dumping, logbooks and catches. Most important to control offshore operations and foreign fleets. The only platform that can effectively conduct an arrest offshore. High deterrence factor.	High cost and limited area surveillance capability. Low rate of detection of infringements.
Patrol planes	Can provide high coverage for identification of illegal incursion of unlicensed vessels and effectively patrol borders and closed areas.	No ability to arrest or to inspect catch or gear.
Helicopters	Can cover relatively large area, can deploy inspectors on vessels and arrest.	High cost and limited distance covered compared to patrol plane
Observers	Can monitor all operations onboard a specific vessel and verify catches, discard, dumping, gear and validation of required documents	Medium cost. Only viable on larger vessels. The integrity of observers may be a relevant question in terms of the quality of data provided.
VMS	Provides up to real time monitoring for licensed or fitted vessels and can reduce interception times for enforcement craft. Low to medium capital and running costs (ship unit bought by fishers)	No coverage of vessels not fitted with the required equipment. Requires integration with other platforms or sensors to be utilized effectively. Technical maintenance and IT support can be limited in some countries.
Satellite Imagery	Full coverage of area scanned	Expensive for regular scans. No positive identification of targets unless verified by other sensors.
Beach Patrols	Efficient tool within recreational and near shore fisheries. Contact with fishers.	Visibility of inspectors, access to remote areas can be difficult.
Navy and coastguard	If available can be free to fisheries organisation, if they are in the field they can monitor border violations.	Limited capability – only border violations as limited fishery knowledge.

Component	Advantages	Disadvantages
Catch and transhipment monitoring	Can monitor landed catch and quotas. Has power to arrest in port. Low capital and running costs	No possibility of monitoring vessels that do not call at port. No possibility of monitoring dumping, gear violations or off-shore transhipments. Information is only of fish landed not those discarded and no geo-referenced data.
Market and sales monitoring	Good information source in terms of landed species and market demands	Difficult to trace the origin of the fish.
Export monitoring	Good information source on volume of landed fish in high value fisheries.	Only part of the landed catch may be exported.
Roadblocks and transport monitoring	Good tool against sale and transport of illegally caught fish.	Roadblocks are easily detected and can be avoided.

3.1 Before fishing

Control of fishing vessels or small craft and fishers before fishing trips, at the time of the issue of a licence, through annual frame surveys or through spot checks is a useful and low-cost MCS operation that can facilitate the following:

- the checking of gear and effort control mechanisms (e.g. horsepower and vessel capacity) to ensure that regulations or licence conditions are complied with;
- if illegal gear is detected or shown then it can often be secured so that it is not possible to use it while fishing;
- to gather information for fishery statistics;
- if vessels have already been fishing it may be necessary to determine if any catch is still onboard;
- this pre-fishing interaction with fishers can be very positive and show the seriousness of the organisation; also MCS personnel get hands on experience with the fishing sector;
- it will generate feed-back from the fishers that may give valuable information for planning or fisher intelligence.

Safety at sea can also be controlled if a vessel is inspected at port. This is an issue that the Code of Conduct focuses on in Paragraph 8.4.1. Fishing at sea is the most dangerous occupation in the world. The drive for economical gain in fisheries has resulted in poor safety for many fishers and this is particularly true for vessels not covered by international instruments such as the Standards of Training, Certification and Watchkeeping for Fishing Vessel Personnel (1995) (STCW-F) which refers to vessels over 24 meters or powered by more than 750 kW. It is important for fisheries management authorities to play a large role within this field in cooperation with the maritime authority. The international instruments also set a minimum recommended standard that can be made valid for smaller vessels within national legislations. Two sets of guidelines to improve the design, construction and equipment of fishing vessels were formulated in the 1960s and 1970s, not as a substitute for national laws but to serve as a guide to those concerned with framing national laws and regulations. These publications are under revision by the International Maritime Organisation (IMO) and are the Code of Safety for Fishermen and the Voluntary Guidelines for the Design, Construction and Equipment of Fishing

Vessels. The safety of fishers is in the interest of the fisheries management authority: this responsibility must not be avoided!

3.2 While fishing

Fisheries MCS operations carried out at sea can have an impact as a deterrent or for enforcement of all control measures but generally they are most significant for output and technical controls. It is the only method that allows infringements in relation to logbooks, gear types and catch to be detected on the site of the crime (while fishing). Important information is also collected at sea that can be time, date and position referenced in relation to both activities and catch.

3.2.1 Logbooks

Logbook data (catch, effort, location, environmental parameters, and gear) is completed by fishers during fishing activities. Logbooks usually need to be designed for each fishery and when good co-operation with fishers can be gained they provide valuable information for scientific assessment, catch monitoring, and feedback to the fishers in terms of historical records. The quality of the data in the logbooks may vary and will relate to the management measures applicable (e.g. in an effort controlled fishery, catch and bycatch data are likely to be more accurate than in a catch controlled fishery), control routines (e.g. regular verification by inspectors or observers or the need for daily radio reports) and the fishers perception of the importance of the logbooks.

3.2.2 Patrol vessels

Fisheries patrol vessel is a very broad term for vessels in a variety of sizes with many different configurations and these vessels along with patrol planes and observers are seen as the traditional tool for MCS. The main principle is that a vessel is able to monitor and enforce fisheries legislation on the fishing grounds. The type of fleet to be controlled may vary from artisanal boats to large foreign trawlers. The fleet to be monitored, sea and weather conditions, possible hostile situations etc. will determine the capacity and configuration required for a patrol vessel. Patrol vessels carrying fisheries enforcement officers are often the only way to obtain some vital and legally acceptable evidence of infringements. It is also the principle platform that can conduct an arrest of a vessel at sea (helicopters also have this feature to a limited extent).

Patrol vessels can be costly to buy and to operate, but they are in many ways irreplaceable, therefore efforts must be made to optimise their operations. Patrol vessels are slow platforms covering relatively small areas, so their main purpose would be deterrence due to their low capacity to detect infringements.

3.2.3 Patrol aircraft

There is a wide range of aircraft available with different performance abilities within the range of light aircraft, which are suitable for maritime and fisheries surveillance. Air operations are very useful for surveillance of large areas and can be utilised in trans-boundary, regional and high seas operations. This MCS component is the only one that can provide an overall picture of a large fishing zone within a short timeframe. Correct utilisation and appropriate information sharing of aircraft sightings will improve deployment of patrol vessels and observers. Aircraft can also have uses for monitoring such as sightings of fish schools, whales, and reef destruction.

Helicopters are more limited than fixed wing aircraft in terms of monitoring large areas effectively but they have the advantage of being able to hoist personnel to and from a vessel. Helicopters normally are 5-10 times more expensive to operate than a small fixed wing aircraft, and therefore the need for one must be clearly identified. The main objectives of a helicopter in the fisheries surveillance role would be the same as for a fixed wing aircraft with the addition of the ability to carry out fisheries inspections including possible arrest of vessels, if applicable, by fisheries officers. This is particularly relevant when the helicopter is carried by the patrol vessel.

3.2.4 Observer programmes

Observer programmes are the only way to implement and to ensure compliance with certain controls such as bycatch or discard regulations that require continuous monitoring. Observers are also able to collect time, date and position information for activities and catches (including biological data) and through this monitor for area and season restrictions and provide valuable information for the scientific organisation. Observer programmes also contribute to deterrence and can create transparency among fishers.

Observers require training, manuals and suitable equipment and supervision to perform their task adequately. Vessels need to be large enough to accommodate observers and possibilities need to exist to place (in port or via the patrol vessel) observers and remove them from vessels. Observers are generally a low cost option for at-sea monitoring and surveillance that have many advantages such as providing continuous contact with fishers, a high deterrence impact and valuable data collecting. Observers do not have the power of arrest so they are only able to record and report any infringements, not to act.

3.2.5 VMS

A Vessel Monitoring System (VMS) provides real-time position, course and speed (PCS) data through a communication link directly into a base station. This allows operators to follow all licensed activity as it happens. These data are sent from a unit on the vessel to a shore receiving station that then displays the vessels on electronic maps with an accuracy of around 100 meters. Satellite communication such as Inmarsat-C is most commonly used, although VMS can be implemented through a range of communication solutions depending on their respective coverage. Fishing in illegal areas, trans-shipments of fish and transfer of fuel can all be indicated through this system. VMS is a tool to assist in more timely and cost effective monitoring and surveillance of authorised and participating fishers. It also significantly supports the more efficient direction and deployment of patrol vessels and patrol aircraft.

Additional opportunities provided by a VMS include the manual entering of catch and effort data (from logbooks) that can be forwarded through the same system for assisting in management of quotas and stock assessment when timely information is required. VMS also creates a solid safety feature for vessels as their position is known at all times and an emergency function is built into the system. Added benefit for the industry is also possible including the option for improved fleet management and catch information that may be available in a timely manner, facilitating improved marketing possibilities (these can be closely linked to the growing electronic marketing of fish and seafood now available).

The validity of VMS information in court needs to be tested for each country. Legal experiences internationally suggest that additional information such as photographic evidence from a patrol vessel, plane, or observer is needed to secure sufficient evidence in case of violations. It is also important to remember that VMS only monitors those vessels carrying active equipment onboard. It will never detect poachers or unlicensed vessels. VMS is highly valuable to assist with area controls, border controls and fleet separation of a regulated fishing fleet.

VMS can be limiting due to its cost for smaller artisanal or in-shore vessels that can seldom be burdened with the cost of the required vessel units. This has generally limited the use of VMS to larger commercial vessels although a trend towards less expensive units is emerging.

3.2.6 Satellite Imagery

The future may open up for additional remote sensing tools that will primarily be useful to supplement the VMS in terms of detecting unregulated fisheries as well as being able to secure acceptable evidence for prosecution of illegal activities. Satellite images are taken from satellites and include radar or photographic images. Radar images have proven particularly useful as they provide good pictures regardless of cloud coverage or light conditions (day or night), while photographic images are more limited by these conditions.

These pictures can become available to fisheries officers in nearly real time (within 2 hours). As fishing vessels move slowly when they fish, the comparison of VMS and satellite images will highlight the presence of illegal fishers and prompt on the spot response from the fishing authority. Present weaknesses with satellite images are that poor sea conditions reduce the detection capability of the system. It is, however, stated that results from different studies confirm that fishing vessels longer than 35 meters in length can be detected at a 95% probability with radar satellites. A second weakness at the present time is that a picture cannot be ordered as a reaction to an incident as the beam programming requires at least 28 hours advance notice.

The present limitations will certainly be reduced as countries and organisations continue to explore the use of satellite imagery. The European Community, Peru, Norway, Canada, The Maldives and many more are participating in research and pilot programmes in relation to integrating satellite imagery with VMS. Potential savings are directly related to aircraft and patrol vessel costs as their efficiency can be increased significantly as the planning and operational deployment of these units will improve.

3.2.7 Beach patrols

In artisanal or recreational fisheries beach patrols may be required to check for fishing licences, bag limits, size restrictions (of fish), gear restrictions or for gathering information. These patrols can be performed randomly or in some type of planned sampling strategy. They can be performed on foot or in vehicles. They will also provide an important interaction with the artisanal or recreational fisher working from the shore (e.g. beach seines, and pole and line fishing) to allow the transfer of information directly to the fishers.

3.2.8 Navy and coastguard

The navy is normally neither designed, educated or particularly trained for fisheries MCS operations. The organisation can be a valuable asset in the sense of monitoring border violations of unlicensed vessels and assistance during hot pursuit, but it is seldom efficient for monitoring catch or gear controls.

A coastguard is more capable of fisheries protection tasks while usually being less advanced than a navy in terms of training and equipment. A coastguard is normally designed around the UN Convention on the Law of the Sea with basic police tasks to perform with emphasis on border violations, fisheries enforcement, search and rescue operations, custom and immigration tasks. It is thus important to remember that any deviation from a pure fisheries protection is a compromise that will reduce the effectiveness of each individual function.

3.3 During landing

The place of landing whether it is a small landing site or a large port provides a bottleneck in fishing operations where vessels can be checked, documents such as logbooks collected and the fish being landed can be identified and weighed. Monitoring of landings is one of the most important elements of MCS operations when output controls are in place. Landing controls are normally less expensive than use of classical MCS platforms as inspectors will be able to travel by road to most ports or landing places, and sampling systems can be developed to suit the local conditions. It is important to remember that monitoring of landings does not detect discarded or trans-shipped fish or fish sold prior to landing. Only physically landed fish can be monitored without knowing where or how the fish has been caught.

3.4 Post landing

Control measures of trade units dealing with fish may be another valuable site where catch data can be verified. Inspections of fish markets, transport providers and sales organisations can provide valuable information about the catches. This type of operation generates valuable information for biological and economical crosschecks as well as validation of other MCS information. It is also a viable operation for control of illegal fish, especially undersized and protected species in general. This is especially valid in small-scale and semi-commercial domestic fisheries where high value catches such as lobster, tuna, sharks and swordfish are caught. Road blocks are another method that can be useful for recreational fishers when bag limits or requirements for licences can be checked.

4. FACILITATING FOR MCS

To facilitate for MCS means that apart from the core MCS system further arrangements or actions are taken to make MCS operations easier, more efficient and more cost effective. Often relatively simple arrangements can result in substantial improvements in the MCS solution.

4.1 Administrative options

Vessel marking system

A proper vessel identification system must be in place in order that patrol vessel crew, airborne personnel or inspectors are able to identify fishing vessels and verify legal vessels effectively. Small registration marks or hand painted registration numbers will make the job of the enforcement units almost impossible. FAO sets standards for marking of fishing vessels that have both proven adequate and provide an easy system to follow (FAO Standard Specifications for the Marking and Identification of Fishing Vessels). These standards are recommended and they support the guidelines given in Article III Paragraph 6 of the FAO Compliance Agreement and Paragraph 8.2.3 of the Code of Conduct. When Malaysia decided to implement a vessel marking system they adopted FAO's standards and in line with these designed a tamperproof registration mark and zonation system. As an incentive to encourage fishers to register they gave a certain timeframe during which registration would be free, and after which time it would be expensive to register. Following this system Malaysia was able to successfully implement a registration and marking system in a few years.

Banning of certain transhipments

Banning of transhipments at sea or outside the port limits is an option to centralise the transhipment operations and therefore make inspections more practicably possible. This option can be applied to certain valuable fisheries or possibly to foreign fleets or across the board to all fishers. It also facilitates the deployment of at-sea observers onto vessels.

Briefing and vessel clearing

In a commercial fishery it can be very valuable to have the captain report to the fisheries authority at the start of each fishing season in order to be briefed about the conditions of a license and to offer the opportunity to collect documentation including logbooks, licences etc. At the same time the vessel will be available for clearing by inspectors.

Checkpoints for vessels leaving a zone

If a foreign fleet is operating then the establishment of checkpoints at certain positions may be a useful system to allow inspections of the vessels before they leave the zone to offload in a foreign port. Alternatively if catch control is very important, offloading of catch can be restricted to specific domestic harbours to ensure complete control of a specific fishery. In Norway vessels have to report at a checkpoint before they leave the Norwegian EEZ for foreign ports. The inspectors are then able to decide if they will inspect the vessel or not: the compliance advantage of this is that the fishing vessels are always prepared for inspection whether it is conducted or not.

Limitation of landing sites

For artisanal or small-scale fisheries many countries have thousands of landing sites making it impossible to control or even realistically to sample landings. One option is to limit the landing sites for a particularly valuable, protected or overexploited fishery or fish species. This channelling of landings to only a restricted number of landing sites makes it easier to deploy inspectors or data collectors to sample the fishery. It may be necessary to support this with spot-checks at other landing sites and markets to ensure a deterrence against violating this regulation.

Special courts

Many courts are not familiar with fisheries violations; this often results in low fines or lost court cases. It may well be worthwhile considering an educational programme or seminars that focus on fisheries legislation and related infringements for court staff and judges, stressing the importance of fisheries management to the country, the possible economical gains that illegal fishing has for offenders and the requirement and effect of high deterrence. Alternatively it is possible to train and allocate special judges or courts to handle fisheries violations.

Coastal zone separation

Conflict between fishers using different gears, between small-scale and large commercial fishing vessels or different resource users is a common problem for an MCS organisation. MCS solutions in a coastal fishery can be very complex due to the multiplicity of resource users and the often difficult access to the resource and landing places. Finding the appropriate solution will need involvement with other managers in the coastal zone and possibly involve the consideration of zones for different users or fishing types. Such options may contribute significantly to decrease tensions and conflicts between the different fisheries or participants and often facilitate self-regulation.

Joint committees

It is important to consult all interested parties when designing and implementing the MCS solution as discussed in Section 2.2.8 and Chapter 7 of the Guidebook. In order to facilitate this participation, joint committees can be formed that meet on a regular basis to allow dialogue and exchange of information. This type of co-operation is valuable within all fisheries ranging from artisanal to large commercial fisheries.

4.2 Information management and sharing

A MCS system will produce large amounts of information through the different monitoring and surveillance programmes. Some of this information is required almost immediately for surveillance activities and to co-ordinate the effective deployment of MCS components, while other information is needed in a less timely manner but over a longer time-series. These different requirements for information make good information management vital. The definition of 'good' is not an easy one: striving for accurate and timely information is important, but also the concerns of what information and in what format are a vital question. It is far too easy to collect too much information, which then is a burden on administration and database systems to compile, check, retrieve and store. The need for information should determine how much and what information is collected, how it is compiled, shared and processed.

It is important to consider both information sharing inside MCS and dissemination of the outputs. Even if exact data requirements differ between MCS, scientists or economists there is a large overlap and the sharing of information is important. The MCS organisation could for example improve deployment of inspectors, observers, patrol vessels and planes if seasonal statistics within the different fisheries were shared between the scientific and surveillance personnel. Stock assessment often suffers from a lack of basic data that can often be enhanced by observer, data collectors' or inspectors' information. Economical data such as catch value and market data will give an indication of the financial status the fishing fleet is operating under: the probability of infringements and illegal fisheries generally increases with reduced profit and difficult market situations. The market situation will also influence the fishers' perception of what is fair and not fair in terms of management measures and resource rent and consequently have an impact on voluntary compliance: a difficult situation for the MCS manager but one that a flexible strategy will be able to deal with if adaptive deployment is possible and the information on markets is readily available. There are many more examples, but the important message is that information sharing between the different areas within fisheries management is vital for the optimal development of the fisheries administration and MCS system.

It is not essential to have an electronic system for information management but it is often the ideal choice if the infrastructure and personnel can support it. If it is opted for, it is important not to become too ambitious and to plan the system within the capability of the organisation both financially and in terms of personnel skills; this is especially relevant when the previous systems have been manual. Implementation and the training of personnel to maintain a newly computerised system will take time and a sensible approach is to design and implement a phased system with one or two aspects of the system being implemented at a time, with integration and linkage occurring later. It is also recommended to adopt a standard software development life cycle even for the development of small systems; information on these is available in any basic reference material for systems development.

4.3 Management system

While the policy level aspects of MCS are normally (and should be) firmly rooted in government, the operational aspects need not be. Conventionally the fisheries management authority, the navy, the police or the coastguard have been responsible for the operational aspects of MCS systems.

For cost and efficiency reasons, contractually engaged private MCS operators are gaining acceptance; examples include the observer programmes of Canada, US and Australia, air surveillance in Canada, patrol planes and certain coastguard vessels and fisheries protection vessels in Norway. The question to be asked is which functions of the MCS organisation are core functions and which are non-core functions. For example the core functions of designing strategies, fining and arresting violators, interpreting VMS output and observer violation reports are usually kept within the fishery management authority itself. However, why should the MCS organisation train pilots, engineers and have the expenses of operating an aeroplane when all that is needed is 8 hours of air patrol per week? The function of flying the plane is not a core function and it may be more viable to provide a fisheries officer to join a private aeroplane tasked for the required flying time.

This option to outsource or privatise non-core functions is an option that should be explored within the fields of MCS in line with the legislative framework for fisheries. It is often more efficient and more cost-effective to rent necessary services from a private operator and thus to remove the burden of maintenance and training of technical personnel and equipment maintenance from the core organisation.

5. ENSURING SYSTEM PERFORMANCE

Measuring performance of the MCS system against the strategic targets should be an annual activity of the MCS organisation and it should involve feed-back from involved interested parties. It is a fact that the perfect MCS system with 100% compliance does not exist. It is therefore imperative to explore what level of compliance is required when a MCS strategy is developed and to compare this to the actual level of compliance being achieved. These actions will also encourage the scientists to use these parameters as a part of their calculations for stock assessment. The definition of the level of compliance required will depend on two main points; an evaluation of the risk related to the sustainability of the stock on the lower limit, and an evaluation of the cost factor for the upper limit.

5.1 Assessing MCS performance

MCS performance is not measured by the number of arrests or prosecutions (legal actions) as this does not reflect the level of compliance, which is the true measure of a successful MCS system. The most practical way to estimate compliance is to compare the number of detected infringements in relation to the percentage of the population being sampled (vessels, fishers, gears etc). The number of infringements can then be raised to the estimated number in the entire population that is being sampled on a monthly, seasonal or annual basis. The resultant estimate may not be completely accurate for many reasons but it does give a reasonable estimate of the level of compliance for a given management measure and can be compared both to the target and across time as a trend for changes in compliance.

An important factor to remember is that a high level of compliance or an improvement in compliance over time is a better measure of success in the system than a higher number of detected infringements. However, assessing the number of apprehensions or convictions, the number of observer sea days and fish sampled or the number of inspections and patrols will also provide details and statistics on the effort exerted in the MCS system that are also important for annual planning and control.

The following questions should be frequently asked to assist in measuring the effectiveness of the MCS system.

- What are the goals and objectives of compliance in the different fisheries?
- What were the expectations of the system?
- Are all MCS strategies implemented in an effective and efficient way?
- Are there changes in the fishing fleet or within certain fisheries that are not covered under the present MCS operations?
- Is there new technology or other means that can improve the MCS system?
- Do the fishers accept and comply with the fisheries legislation (if not find out why)?
- Are the staff performing as expected (if not find out why – it could be lack of resources, management skills, training, support)?

If these questions are regularly considered and analysed they will give some indication of the optimal levels of performance aimed for and the answers may help in accessing the best way to achieve the required outputs. Performance is assessed against the original objectives, and these will be a reflection of the overall fishery management objectives and the contribution of the MCS system towards these must be considered.

5.2 Cost analysis

The chances of detecting violations of fisheries regulations are directly related to the amount of resources used for monitoring and surveillance, and how efficiently these resources are used. Therefore cost-benefit analysis is required and this is closely linked to the assessment of performance. An increased level of compliance should be related to either an increased income to the State or decreased biological risk (which ultimately will be a financial gain to the State). The desired level of compliance will to a large extent determine the cost or visa versa. This is again related to the type of fishery, state of the stock, the value of the fishery and geographical considerations (e.g. number of landing sites, sea conditions, number of ports etc.).

Cost analysis in relation to performance must be prepared both historically and for the future. Essential questions are whether the resources applied are giving the desired results and if other components would achieve better results within the financial framework. This must particularly be kept in mind when costly hardware is considered: a new and larger patrol vessel will normally have a life cycle of 20-30 years with annual cost implications. Cheaper solutions like leasing a vessel or using observers may be a more sensible solution if the future of the fishery is uncertain.

6. CONCLUSION

MCS has historically often been perceived as a somewhat isolated element of fisheries management dealing primarily with the enforcement of legislation. This chapter has revealed that MCS in fact relates to all of the activities performed by a fisheries management authority in relation to the actual fishing operations. Setting effective fishery management strategies requires an integrated approach with a full understanding of the needs and constraints of the management system including those that an MCS organisation has in implementing the management measures. On the other hand the MCS organisation must understand the principles of fisheries management to be able to carry out their operations and in order to contribute useful information to the management process. Old barriers between the different components of fisheries management have to be removed to create a successful integrated fisheries management regime.

MCS is not intended to be a policing function where fishers are treated as criminals. The prime function of MCS is to increase compliance to agreed management measures by increasing deterrence and voluntary compliance and thus decreasing violations. A strategic balance between the two aspects of deterrence (or enforcement) and voluntary compliance should be achieved for each fishery. Participatory management involving fishers and other interested parties is seen as a key tool required by all fisheries (artisanal, small-scale and commercial) in order to ensure an increase in compliance.

Many MCS solutions exist for a given fishery: selecting and compiling the components in the most cost-effective manner is not an easy task. The value of cross verification and achieving a balance between the different dimensions (before fishing, while fishing, during landing and post landings) has been introduced as an important element to consider when designing the MCS solution. This will not only provide the organisation with several different sources of information but also increase the overview of the fishing sector both from an information and deterrence point of view. The desired and expected level of compliance, the value of the fishery and the state of the stock(s) are all important factors that help the fishery manager to establish priorities to assist in allocating the MCS resources between components.

Large complicated fisheries often demand complex MCS solutions: these can often be assisted through simple limitations in terms of vessel marking, transhipment and landing sites. These options and more have been introduced to the fishery manager in order to assist in the facilitation for MCS, while, low cost and low technological options have been introduced for artisanal and small-scale fisheries with limited human or financial resources.

Finally, every MCS system requires regular assessment in order to ascertain if it is achieving the strategic targets in the most cost effective and efficient manner. This chapter has described the key element of assessing system performance to be a comparison of the levels of compliance over time and against the targets set out in the MCS strategy for given management measures. It is suggested that an improvement in compliance over time is the indication of a successful MCS organisation and system.

7. RECOMMENDED READING

Davies, S.L. 2002. Guidelines for developing an at-sea fishery observer programme. *FAO Fish.Tech.Pap.*, **414**. FAO, Rome. 118 pp. (in preparation)

FAO. 1997. Fisheries management. *FAO Technical Guidelines for Responsible Fisheries.*No. 4. FAO, Rome. 82 pp.

FAO. 1998. Fishing Operations, 1. Vessel monitoring systems. *FAO Technical Guidelines for Responsible Fisheries.* No 1. Supplement 1. FAO, Rome. 58 pp.

FAO. 1981. Report on an expert consultation on monitoring, control and surveillance systems for fisheries management. Rome, 27-30 April 1981. FAO/GCP/INT/344/NOR. FAO, Rome. 115 pp.

FAO. 2000. *The State of World Fisheries and Aquaculture.* FAO, Rome. 142 pp.

Flewwelling, P. 1994. An introduction to monitoring, control and surveillance systems for capture fisheries. *FAO Fish.Tech.Pap.*, **338**. FAO, Rome. 217 pp.

Hersoug, B & O. Paulsen. 1996. Monitoring, control and surveillance in fisheries management. University of Namibia, Windhoek, Namibia. 378 pp.

Kuperan, K., & J. G. Sutinen. 1998. Blue water crime: legitimacy, deterrence and compliance in fisheries. *Law and Society Review*, **32**(2): 309-338.

Laurec, A. 1999. Monitoring fisheries for better research and/or better enforcement. *Proceedings of the International Conference on Integrated Fisheries Monitoring.* Sydney, Australia, 1-5 February 1999. FAO, Rome. 378 pp.

Smith, A.R. 1999. *Monitoring, control and surveillance in developing countries and the role of FAO. Proceedings of the International Conference on Integrated Fisheries Monitoring.* Sydney, Australia, 1-5 February 1999. FAO, Rome. 378 pp.

CHAPTER 9

DESIGN AND IMPLEMENTATION OF MANAGEMENT PLANS

by

David DIE

Rosenstiel School of Marine and Atmospheric Science, Miami, U.S.A.

1. INTRODUCTION 206
2. DESIGNING A MANAGEMENT PLAN 206
   2.1 What should a management plan contain? 206
   2.2 International fisheries policy requirements 207
   2.3 National/State fisheries policy requirements 207
   2.4 Fishery-specific requirements 208
   2.5 Who should design a management plan? 208
   2.6 Timetable for developing a management plan 209
3. IMPLEMENTATION OF MANAGEMENT PLANS 210
4. REVIEWING AND AMENDING MANAGEMENT PLANS 211
   4.1 Mechanisms for review 211
   4.2 Review strategy 211
   4.3 Changing management measures without amending the FMP 212
5. FISHERY MANAGEMENT PLANS WITHIN THE CONTEXT OF ECOLOGICALLY SUSTAINABLE DEVELOPMENT (ESD) 212
6. EXAMPLES OF MANAGEMENT PLANS 213
   6.1 The Australian Northern Prawn Fishery (NPF) Management Plan 213
   6.2 The Barbados fisheries management plan 213
   6.3 Queen Conch Fishery Management Plan for Puerto Rico and US Virgin Islands 216
   6.4 Western Australia plan for developing new fisheries 217
7. CONCLUSIONS: HOW DOES A MANAGEMENT PLAN HELP TO ACHIEVE THE MANAGEMENT OBJECTIVES OF A FISHERY? 218
8. REFERENCES 219
   8.1 Web resources 219

The building blocks for good fisheries management have been presented in the previous chapters. The first few chapters covered the techniques that can be used to manage fisheries. Chapters 6 and 7 specified how fishing rights can be established to allocate fish resources to certain users and how participation of interested parties in the management process can improve the effectiveness of management. In Chapter 5 the information requirements to support good management were outlined. It is time now to show how everything can be put together into a single instrument that outlines how management is actually going to proceed for a specific fishery. Such an instrument is what is called a fishery management plan (FMP) and is the subject of this chapter.

1. INTRODUCTION

Planning is an essential part of the management process regardless of whether one deals with the management of a fishery or the management of a car assembly line. The management plan is the main instrument that specifies how management is to be conducted in the future. In many fisheries, fishery management plans (FMP) are often also instruments not just for planning but also for operational management. These plans don't just document the way to reach management goals in the future (strategic), but also describe how to manage the fishery in the present (tactical). This dual purpose of fishery management plans is not recognised universally, in fact the only mention of an FMP in the FAO Code of Conduct for Responsible Fisheries appears in Paragraph 7.3.3 that states,

> "Long-term management objectives should be translated into management actions, formulated as a fishery management plan or other management framework".

The FAO Technical Guidelines for Responsible Fisheries (FAO, 1997), however, clarify that FMPs have a tactical component to them that defines day to day management:

> "The management plan provides detail on how the fishery is to be managed and by whom. It should include a management procedure which gives details on how management decisions are to be made in accordance to developments within the fishery...".

In fact these FAO guidelines give a very specific definition of an FMP:

> "A fisheries management plan is a formal or informal arrangement between a fishery management authority and interested parties which identifies the partners in the fishery and their respective roles, details the agreed objectives for the fishery and specifies the management rules and regulations which apply to it and provides other details about the fishery which are relevant to the task of the management authority."

In deciding upon a definition that acknowledges formal or informal arrangements, FAO draws attention to the fact that having a formal document describing an FMP may not be the only way to achieve the objectives of management. It acknowledges the fact that in some fisheries there are management arrangements that are successfully achieving the management objectives of specific fisheries but are not formally collated into a document called an FMP. Although such situations are discussed briefly below and in Section 4.1 of Chapter 6, the focus of this chapter remains the description of the process of how to develop a formal FMP. This is done by broadly following the FAO Guidelines (FAO, 1997) regarding the development, implementation and review of FMPs. Also, some examples are presented to highlight how components of those guidelines can be found in the FMPs developed for some important fisheries around the world. The examples are not meant to be comprehensive, but rather serve to highlight how FMPs have been developed and implemented in real fisheries.

2. DESIGNING A MANAGEMENT PLAN

2.1 What should a management plan contain?

The institutional arrangements pertinent to a fishery are essential in defining the contents of its management plan. The international, national and even regional context within which a fishery takes place (Section 9, Chapter 1) will influence the fishery policies and legislation that govern the fishery. In some countries, like the United States of America or Australia, there are specific references to the need to develop management plans in their fisheries legislation (e.g. Queensland, Australia: Fisheries Act; United States: Magnuson-Stevens Act). That same legislation may specify the main sections that a management plan should contain. At a minimum, FMPs should contain:

- a description of the fishery especially its current status and any established user rights:
- the management objectives;
- how these objectives are to be achieved;
- how the plan is to be reviewed and/or appealed; and
- the consultation process for review and appeal.

More details on the exact contents of real FMPs can be found in the examples presented in Section 6 of this Chapter. For details on how to develop operational management objectives refer to Chapter 5. For an in depth presentation of tools used to achieve management objectives refer to Chapters 2 to 4 and Chapter 6. For a discussion on the type of scientific information that can be used to evaluate the biological, ecological, economical and social status of a fishery read Chapter 5, while Chapter 6 presents an in-depth summary of user rights.

2.2 International fisheries policy requirements

Nowadays one of the most basic requirements recognised in FMPs is the adherence to the, internationally sanctioned, United Nations Convention on the Law of the Sea of 10 December 1982 (LOS Convention), the FAO Code of Conduct for Responsible Fisheries and the precautionary approach to management. Such adherence tends to be recognised in the fishery policies of each State or in the statutes of international organizations in charge of coordinating fisheries management. For fisheries in the high seas and for those plans dealing with straddling stocks, the FMPs must clearly specify the international regulatory and institutional context within which the plan is applicable (e.g. UN Fish Stocks Agreement – see Table 2, Chapter 1). In such cases the FMP should have clear links to both the national fishery management policy and to the statutes of the Commission or international organization in charge of coordinating the management of the international fishery. For example the Australian FMP for Southern Bluefin Tuna has as a first objective:

> "... to ensure, by appropriate management and in conjunction with the Commission (the Commission for the Conservation of Southern Bluefin Tuna), the conservation of the stock of southern bluefin tuna..."

A word of caution must be expressed in reference to national FMPs designed to manage fisheries for resources which, at least partially, are distributed in the high seas, or that are shared between two or more countries. Because these resources are not under the jurisdiction of a single nation the FMP, being a national instrument, will only be applicable within the EEZ of that nation. This does not mean that the FMP will be ineffective, but it does mean that the plan may have to define national objectives that are constrained by the objectives of other countries (see Code of Conduct, Paragraphs 6.12, 7.1.2 and 7.3.2). Chapter 4, particularly Section 5, provides a lengthier discussion on this point.

2.3 National/State fisheries policy requirements

National fishery policies should be supported by a legal and institutional framework so that FMPs can be the main instrument of management. In that respect, every national fishery policy should define the range of minimum requirements that a fishery management plan should fulfil. The national policy should therefore broadly determine the type of information required to be included in an FMP so that all FMPs achieve the goals of the national fishery policy. Because these goals are likely to be broad, the information requirements are likely to be general, however, it remains the responsibility of those preparing each FMP to make sure that the specific objectives of an FMP comply with the national fisheries policy.

2.4 Fishery-specific requirements

The minimum requirements specified by national fishery policies tend to still leave considerable leeway about the contents of an FMP. As a result even within a single fisheries jurisdiction there are large differences between the FMPs developed for different fisheries. This should be considered a strength of the policy, because it ensures that the FMP is developed so as to suit the specific needs of the managers of each fishery and is not just a bureaucratic requirement.

It is essential that the plan specify how the management objectives (operational objectives) of the plan are to be met. If possible each management objective should be examined individually and the management measures that are designed to help achieve that objective should be identified and explained. Here the plan needs to be very specific on how objectives and measures link up and which performance indicators are going to be used to measure the achievement of management objectives. This must be done by structuring the plan in a way that requires the development of performance indicators (Figure 2 of Chapter 1). The indicators must cover all objectives of the plan so in general there will be indicators of the biological status of the stock but also social and economic indicators of the health of the fishery.

These indicators will equally reflect the agreed reference points and, for example, measure how successfully the plan is achieving the target reference points and remaining within the limit reference points. In general, the plan will not specify the exact way the indicators are developed, but it must require their development. As an example, forecasts on the economic impact of fisheries management actions could be used as an indicator of the success of a plan in trying to achieve its objective of maximisation of economic returns. Section 3 of Chapter 5 discusses in detail examples of specific indicators that can be used in FMPs.

The FMP should specifically make reference to the characteristics of the natural environment within which the fishery takes place and to how changes in this environment may affect the management of the fishery. If there are traditional management structures or established rights that have been historically used, these should be recognised and included in the FMP (Chapter 6 discusses user rights).

Management objectives in an FMP often conflict (e.g. resource conservation vs maximising economic returns or promoting development). The plan must acknowledge this conflict and address it through requiring a process by which conflicts between objectives can be resolved, as discussed in Chapter 5. This could be achieved by specifying some of the indicators associated with conservation objectives as constraints, whereas indicators of economic performance might be identified as targets. This will result in the development of operational objectives and will lead to appropriate target and limit reference points. In this context development refers also to human resources, so it encompasses capacity building and enhancement of the quality of life.

2.5 Who should design a management plan?

2.5.1 Institutional arrangements

The development of the FMP is the responsibility of the authority in charge of fishery management, but such authority must make sure that all interested parties in the fishery should participate in its development (Code of Conduct, Paragraph 7.1.2; FAO, 1997). For a more detailed discussion on how to involve community organizations in the management process refer to chapter 7.

2.5.2 Participation of Interested Parties

As discussed in Chapter 7, all interested parties should be offered the opportunity to participate in the development of an FMP. The identification and consultation with interested parties should be one of the first steps to be conducted in the development of an FMP. The earlier the interested parties are involved in this development the greater the sense of ownership of the final FMP they will have. By participating in the process they will be more aware of their rights and responsibilities towards resource management and will tend to comply better with management provisions. All steps in the consultation with interested parties (e.g. comments on discussion paper) should be formally structured and described within the FMP document. In cases were there are substantial differences in the capacity of interested parties to participate in this process, the plan should include capacity building. Chapter 6 (Section 3) discusses further the concept of "management rights" the right to participate in the management process and Section 7 of Chapter 7 details equity issues related to the capacity of different types of interested parties to participate in the management process.

Increasingly there is a requirement to attempt to coordinate the management plans of all marine resources, including fisheries. Other agencies are developing their own marine resource management plans and fishery managers increasingly recognise the need to influence the management of impacts on the ecosystem and protect those habitats and resources that are critical for the health of the fishery. This need to coordinate management leads to having to make sure that the groups included in the public consultation process include all management agencies that have responsibility for management of activities or resources that may be related to or affect fishery resources or to the environment that supports them (Code of Conduct, Paragraphs 10.1.2 and 10.1.5). Because these agencies are often developing management plans themselves this process of consultation can be difficult and lengthy. Often other marine management agencies have different perceptions about the importance of resources within the marine environment to those of the fishery agencies. This often creates differences between what conservation agencies think is achievable and what the fishery agency thinks is achievable. One way to resolve these differences is through well-structured negotiations established within the consultation process for the FMP.

2.5.3 Expert knowledge

The development of an FMP requires extensive information about the fishery and the social, economic and natural environments within which the fishery operates. The gathering of information, in the form of data or expert knowledge, is the responsibility of the management authority. More detail on which information should be collected and how it should be presented to those in charge of the development of the FMP is discussed at length in Chapter 5.

2.6 Timetable for developing a management plan

In theory, the development of a management plan is the first management action that any responsible fishery agency should take when it starts managing a fishery (FAO, 1997). In practice, however, management plans were only recently developed for even the most important fisheries of the world. The first step in the development of an FMP is usually to create a working group that develops a discussion paper. Public comments are then sought through meetings and letters and analysed and formally responded to. Next a draft management plan is developed and released. Public comments are again sought, analysed and formally responded to. After that the final management plan is developed and sent to the appropriate minister for approval before coming into effect. Below more details of each of these steps can be found.

2.6.1 Discussion Paper

When an FMP is about to be developed, it is best to put it out as an initial document for public comment that describes, in layman's language, the reasons for developing the plan and the proposed contents of the plan. This document is similar to that of an FMP but must be easy to read and often will contain more background fishery information than the FMP. This document –often referred to as a discussion paper – should aim at two things, informing the public and interested parties about the plan and seeking their comments. Important information such as current legislation or summaries of the knowledge about the status of the fishery should be included as appendices to the discussion paper. These documents can have a list of specific questions attached to each of the major issues identified by the group developing the FMP. These questions are often part of a formal questionnaire included with the discussion paper and designed to elicit comments by interested parties. General comments, other than those sought through the questions, should also be solicited in the questionnaire. It is important that the management authority encourages representatives of all interested parties to take time to review the discussion paper and make their comments to this authority. Participation of all interested parties is essential and should be facilitated by the management authority as much as possible. This is especially important because the group of people involved in the initial development of the discussion paper have the opportunity and responsibility to define the scope of the final plan. Initial investment in the consultation process will save a lot of resources later and will help the plan to have the highest possible initial acceptance when it is implemented.

2.6.2 Public consultation

Once the discussion paper is formally released, the public and interested parties should be given a set time to make comments, for example three months. During this time it is advisable to organise meetings between the management authority and interested parties (e.g. meetings in the major fishing ports) to seek comment and promote discussion. Transparency during this process will later help during the implementation of the plan. After the comments have been collated the management authority should formally answer these and revise whichever sections of the plan the authority deems appropriate.

The public consultation period is also a period of negotiation and the effort to be put by the developers of the plan in this process should not be underestimated. This is due to the fact that consultation inevitably leads to the presentation of opposing views about the management process. These opposing views must always be measured against the management objectives established for the plan and the national fisheries policy. The fishery management authority must therefore make sure that the special interests of some interested parties do not override the principles contained in the national fishery policy.

2.6.3 Draft Management plan

The first draft management plan is then released and new comments are sought, again specifying a time frame of a few months. The draft management plan is likely to be quite a different document to the discussion paper especially because of the legal language that is to be included in certain sections. It is therefore harder for people to understand and comment on this document. This highlights the benefits of releasing initially a discussion paper. The fishery management authority should, again, devote attention to ensuring that the draft management plan is understood and well-accepted by the interested parties. After the second period of comment is finished, and unless there are extraordinary circumstances that require another review of the plan and a third round of consultation, the final plan draft should be prepared and submitted to the ministry of fisheries, or equivalent, for approval. To ensure that delays in this process do not undermine the capacity of the fisheries authority to conduct good management, it should be at all times emphasized to the public and interested parties that FMPs are to be reviewed periodically and that it is not precautionary to delay necessary management actions when the status of resources requires conservation measures. The fisheries ministry must also be firm in this respect and should avoid political interference getting in the way of the implementation of a plan that has been developed through the appropriate consultation process.

3. IMPLEMENTATION OF MANAGEMENT PLANS

Once the FMP has been approved it is important to inform the public of its contents. A good strategy is to summarise the major points of the plan in easy-to-read leaflets or brief documents that can be distributed to interested parties. These documents will serve the purpose of informing the public of the contents of the new plan and are likely to generate comments from the public that can be considered during future reviews of the plan. Hopefully these documents will also ensure greater compliance with the regulations by explaining the reasons why these regulations are in place.

Management plans have to consider the likelihood that they will be complied with and the enforcing requirements to ensure such compliance (Code of Conduct, Paragraphs 7.7.2 and 7.7.3). A management plan that cannot be properly enforced may damage the credibility of the management authority and therefore undermine the management of other fisheries (FAO, 1997). It is essential that during the development of the plan, fishers and other interested parties affected by plan rules are asked whether these rules are likely to be complied with (see Section 2.2.8 of Chapter 8). The plan should also emphasize the consequences of non-compliance and may often include a description of the penalties (loss of license, fines) when serious offences occur. In addition to the details on monitoring, control and surveillance provided in Chapter 8, Chapter 6 specifies how to implement a system of user rights, and how such system can help compliance. Chapter 7 describes how community groups can facilitate the implementation of the plan and help in the compliance of plan rules.

4. REVIEWING AND AMENDING MANAGEMENT PLANS

Factors of importance to fisheries change through time; therefore FMPs must be periodically reviewed (see Figure 10 in Chapter 5, and Chapter 4). If possible, during the development of the management plan, such changes should be predicted and included in the section of the plan that defines the review process. For example if it is known that a new management measure (e.g. establishment of a marine reserve) may take several years to have a detectable effect, the plan can call for a review of the measure after the required number of years have passed.

4.1 Mechanisms for review

The mechanism for review should be specified in the plan itself. In general the consultation process should parallel the initial process used to develop the plan, but is likely to be shorter in time and should only require one draft review document seeking comments from interested parties and the public, and a final draft to be submitted for approval. Major reviews may require public meetings where interested parties can air their views about the proposed amendments to the plan. Because of the need for public consultation, whenever possible, it is best to break down amendments to the plan into small and discrete components, rather than attempt to change all the shortcomings of the FMP in a single amendment. This strategy has been successfully used in the United States of America to modify management plans of complex multi-species fisheries such as the Gulf of Mexico Reef fish fishery or the South Atlantic Snapper-grouper fishery. In both of these fisheries an amendment has been prepared almost every year by the respective management councils. Some of these amendments are developed and approved within a few months but others may take longer. Sometimes the councils are considering more than one amendment at a time. By breaking down the process of review into small steps the management councils are successfully and continuously improving the plan. Chapter 7 details processes that can help the review process at community level, and provides useful concepts on how reviews may be implemented in artisanal fisheries.

4.2 Review strategy

Often, plan reviews are motivated by changes in the socio-economic status of the fishery or the biological status of the fish stock. It is to be expected that, after the initial development of a plan, it will take several years to close the information gaps that may have been identified at the time the plan was developed. Therefore a major review of an FMP is unlikely to happen until several years have passed and researchers and managers have had time to review and evaluate the need for and effects of possible new management regulations. It must be remembered, however, that the lack of scientific information cannot be used as an excuse for inaction and that the precautionary approach calls for management action on the basis of the best information available (Code of Conduct, Paragraph 7.5.3). FMPs therefore must be reviewed whenever it is precautionary to review the plan, not just when new data become available. It is therefore recommended that within the FMP a regular schedule for reviews is defined. At a minimum, a plan should be reviewed every five years.

4.3 Changing management measures without amending the FMP

In some instances the process of review of a management plan takes too long for it to be an efficient way to make an urgent change of a management measure. This is often due to the fact that FMPs are often legislated documents that require a specific and lengthy procedure to change them. It is therefore advisable to build within the plan the facility to make changes that do not require amending the plan. In Australia, for example, regulations that have to change at short notice such as the start and end of the fishing season or annual changes to a TAC are introduced through executive rules (in Australia these are referred as "directions") by the fishery management agency. These rules have legal standing but that do not require amendment to the plan. This is achieved by defining, within the FMP, the nature and conditions by which these rules can be introduced.

5. FISHERY MANAGEMENT PLANS WITHIN THE CONTEXT OF ECOLOGICALLY SUSTAINABLE DEVELOPMENT (ESD)

Initially, fishery management was limited to the control of harvesting the target resource, without consideration of any effects of harvesting on other resources. Later, management of bycatch species was introduced into the fishery manager's agenda. Finally, the management of indirect impacts on other marine species that depend on fishery resources as a source of food and the impact of fishing gear on marine habitat have now become an important part of fishery management (e.g. Code of Conduct, Paragraph 7.2.2.). All these issues must be considered during the development of a FMP. Section 5 in Chapter 2 considers the impact of different fishing gears on the broader ecosystem and Section 2.2 in Chapter 3 discusses ways to manage these impacts through the use of area and time restrictions.

In many countries, it is accepted that broad ESD considerations must be part of fishery policy. In practice, however, this has not always translated into operational changes to FMPs, partially because most issues relating to fishing impacts on non-target species or marine habitat are poorly understood and studied. There tends also to be few management measures that can alleviate negative impacts on the environment, e.g. habitat destruction produced by trawls, without severely disrupting fishing operations.

This lack of information does not mean that ESD issues can be ignored during the development of an FMP. It is essential that these issues be at least identified within the plan. If there is no information to quantify the importance of the impact or on the capacity of managers to control it, the plan should at least specify how such information will be obtained in the future and provide a timetable of actions to gather such information (See Article 12 of the Code of Conduct and Section 5 in Chapter 5).

In cases where the management of the coastal environment or non-fisheries marine resources is not the responsibility of the fisheries authority, it is essential to link the FMP to other management plans such as coastal zone management plans (Code of Conduct, Paragraph 6.9). At the minimum the FMP should clearly define the agencies responsible for management of the coastal and marine environment that may be affected or have an interest in the FMP. Ideally, the FMP should be developed to help reach the objectives of coastal zone management plans for the areas where the fishery is taking place. This might be achieved directly with instruments of management that are available through fishery legislation, such as prohibiting the cutting of mangroves or damage to seagrass beds. Fishery agencies must be part of coastal area management processes in order to make sure that, when appropriate, the FMP becomes one more instrument to achieve the objectives of these processes. This will also help to ensure that other sectoral management plans developed within these processes help to achieve the objectives of the FMP. Section 3.2 in Chapter 3 discusses some of the difficulties found in integrating fishery objectives into the multiple-use framework of coastal area management.

6. EXAMPLES OF MANAGEMENT PLANS

In previous sections the overall framework of how to develop and implement an FMP has been presented. Four examples of existing FMPs are now briefly presented to help to put this in the context of current fisheries management. These examples were chosen to represent a wide spectrum of existing FMPs. The first example is for a single industrial fishery in a developed country; in contrast the second example corresponds to a mixture of industrial and artisanal fisheries in a developing country. The third example refers to a plan for a single species caught as part of a multi-species artisanal fishery. The final plan is meant to demonstrate how to plan for new developing fisheries.

6.1 The Australian Northern Prawn Fishery (NPF) Management Plan

The NPF is a fishery with only one species group as target, where the only gear used is the trawl and that operates offshore of a very remote part of the world, in Northern Australia. Although the FMP was only developed in 1995, the NPF has been closely managed since the 1980s. This plan is therefore an example of an FMP for a well-managed industrial fishery. This plan was made in accordance with the Australian Fisheries Management Act of 1991. Its purpose is to make sure that the policy objectives of the Australian Fish Management Authority are met in the Northern Prawn Fishery (NPF) and that bycatch in this fishery is reduced to a minimum[1]. This is translated in the plan by specifying that the objective of the plan is "ensuring that the exploitation of fisheries resources and the carrying on of any related activities are conducted in a manner consistent with the principles of ecologically sustainable development and the exercise of the precautionary principle, in particular the need to have regard to the impact of fishing activities on non-target species and the long term sustainability of the marine environment; and maximising economic efficiency in the exploitation of fisheries resources" (Anon., 1995).

This plan is made under clear guidelines established in the Australian Fisheries Policy, and as a result the plan itself is limited to a description of the operational management details for the fishery and does not cover general fisheries policy.

The FMP starts with a list of legal definitions of terms that are used throughout the plan (Table 1). The next section covers the objectives, management measures and performance measures or indicators. Because the NPF is a fishery managed by input controls (see Chapter 4), including limited licenses, section two of the plan focuses on fishing rights (see Chapter 6). The section covers the types of rights that exist, how they are to be transferred and the obligations of fishing rights holders. The last section of the plan contains a detailed description of the managed area and a list of all amendments.

6.2 The Barbados fisheries management plan

The Barbados Fisheries Act (1993-96) required the Chief Fisheries Officer to develop a management plan for the fisheries of Barbados. In 1997 the Fisheries Advisory Committee, in consultation with the fishing industry and the general public, completed the FMP (Anonymous, 1997). Although the fisheries of Barbados, like those of many other developing countries, are

---

[1] Note how the objectives refer to those from an established national fishery policy, but also how special consideration is given to an ESD issue: the reduction of bycatch.

highly diverse, the government decided to develop a single management plan for all of them. This contrasts with many other countries where fisheries management plans are developed for individual fisheries. As a result, the fisheries management plan for Barbados has much broader goals than those found in other plans. These goals appear at the beginning of the FMP document (Table 2) and include meeting human-nutrition, social and economic needs, whilst integrating fisheries policy within coastal zone management and considering traditional knowledge of fisheries and the special interests of local (coastal) communities. Other goals of the Barbados FMP are more commonly seen in other plans, such as maintaining or restoring populations to the levels that can produce maximum sustainable yields, promoting the use of selective fishing gear to minimize wastage and bycatch, researching, monitoring and controlling fishing operations and fish resources, protecting endangered species and fragile ecosystems and finally cooperating with other nations in the management of shared, straddling and migratory stocks. The plan then contains an overview of the fishing industry which, obviously, includes the whole variety of fishing practices and resources found in the country: from shallow water trapping for reef fish and lobsters to oceanic gillnets for flyingfish, handlines and longlines for coastal and oceanic pelagics and hand gathering of sea urchins.

Table 1: Outline of the Australian Northern Prawn Fishery Management Plan of 1995

Part 1 *Introductory Provisions*

1. Name of Plan
2. Commencement
3. Interpretation
4. Objectives
5. Measures
6. Performance criteria

Part 2 *Statutory Fishing rights*

7. Gear statutory fishing rights
8-13. Types of statutory fishing rights (fishing licenses)
14-17. Who may fish in the NPF area
19. Boat nomination and replacement
20-24. Cancellation of statutory fishing rights
25. Directions by AFMA (length of fishing season etc.)
26. Transfer of statutory fishing rights
27. Expiry of statutory fishing rights

Part 3 *Miscellaneous*

28-35. Certificates, delegation, leasing arrangements of statutory fishing rights

Schedule 1   Area of the Northern Prawn Fishery

There is then a description of the fisheries management process used to develop and implement the FMP, and the need to link the FMP to the coastal zone management plan is identified[2]. The plan then describes the legislation that directly influences the plan, and includes a history of previous and existing bilateral fishing agreements with other nations. The next section of the plan defines the organizational framework of the fisheries sector in Barbados, including government and non-government fisheries related organizations and any fisheries programs administered by international organizations. The section ends with a description of the research, monitoring, surveillance, licensing and inspection activities conducted in Barbados.

Table 2: Outline of the Barbados Fisheries Management Plan

Guiding Principles (mission, goals, fisheries policy and country profile)

Fishing industry profile (overview of fisheries, fishing industry, intersectorial linkages)

Fisheries Management (fisheries planning process, coastal zone management, fisheries legislation, regional fishery agreements, organizational framework, research and statistics, monitoring control and surveillance, inspection, registration and licensing)

Fisheries Development (Visions from harvest, postharvest and State sectors)

Management and implementation for specific fisheries (one for each fishery)

Fishery management options

Glossary

The plan then presents an analysis of issues of importance to the harvest, postharvest and government sector. For each issue a series of optional management actions are identified and implementation strategies are proposed, including a description of resources required. An example of an important issue for the harvest sector is the lack of fisher and boat owner organizations. Possible actions to address this issue are to promote organizational development, provide incentives and training. Strategies to achieve these actions are for example to subsidise certain organizations and provide extension training in organizational development. Of course the plan notes that to carry these out, funds and trainers will be required. Although the goals of the plan are broad, an in-depth analysis of all issues allows the government to address them one by one within the priority order established by the policies of the government of Barbados and as a function of the resources available for its implementation. It is expected that as some of these issues are resolved they would disappear from future versions of the plan. Again the plan is a living document.

The final part of the FMP includes sub-plans for each of the eight major fisheries of Barbados. These sub-plans are brief, 2 to 3 pages long, and include concise descriptions of the target species, bycatch, ecology, fishery, management unit, resource status, catch and effort trends, specific management policies, objectives and approaches already in place for such fisheries and a list of development opportunities and constraints. This descriptive part is followed, as in the main part of the FMP, by a list of issues and the proposed actions identified to address them, together with the resources required. At the end the plan includes a list, with non-technical descriptions, of fishery management options used in the FMP and a glossary.

---

[2] Interestingly the FMP acknowledges that the link between the FMP and the coastal zone management plan has not been made, but at least by identifying the need, it highlights its importance. This also shows that FMPs are live documents that may not always have all the issues sorted out before they are adopted.

6.3 Queen Conch Fishery Management Plan for Puerto Rico and US Virgin Islands

The Magnuson-Stevens Act of the United States of America (USA) requires that Regional Fishery Councils develop FMPs for resources within each region. The Caribbean Fishery Management Council developed a plan for the management of queen conch within the waters of the USA Caribbean in 1996 (Anon. 1996). Other similar plans are in effect for Corals and Reef Associated Plants and Invertebrates, Shallow Water Reef Fish, and Spiny Lobster.

Queen conch are caught throughout the Caribbean, where they are a valuable resource for artisanal fishers. The resource seems biologically overfished in many countries, including the USA Caribbean. A fundamental problem of the management of this resource is that the stock is shared by many countries, therefore management may need to be coordinated across several countries. The Queen Conch FMP recognises that explicitly in its executive summary, highlighting the need for both local management actions and regional cooperation.

The FMP starts with a list of definitions of all technical terms used in the document and is followed by an introduction that defines the context (USA fishery legislation and Caribbean Fishery Management Council's management program) and scope of the plan (Table 3). The next two sections are the lengthier part of the plan where all the relevant information on the biology of conch and its fishery is summarised. These sections must contain whatever information is required to show that the best information available has been used to support the management plan. They present details on the life history, population parameters, conch habitats, history of the fishery, fishing fleet, processing sector and the most current assessment of the status of the fishery at the time the plan was developed.

The next chapter of the FMP discusses the most important issues facing the fishery, including the presence of overfishing, the limits on enforcement, the legislative setting, the limitations of current databases, the dependence on communication with and education of interested parties and the importance of habitat quality.

Section five of the FMP starts presenting the management objectives:

1. "to optimise the production of queen conch...
2. to reduce adverse impacts...such as harvesting immature and reproducing individuals....
3. to promote the adoption of functional management measures that are practical and enforceable....and the promotion of international cooperation in management...
4. to generate a data base that will contribute to the knowledge and understanding of queen conch biology....
5. to recommend habitat improvements to federal and local governments...
6. to provide as much flexibility as possible with the management...."

The rest of the section defines why conch has been assessed to be overfished and presents the rebuilding plan to recover the stock. The development of the rebuilding plan is a requirement of the USA Magnuson-Stevens Act.

The plan then details the alternative management measures that are to be used to manage the stock of conch. For each alternative the expected consequences of using the measure are detailed. The alternative of not doing anything is also presented including its consequences. Among the measures included in this list are size limits, sale restrictions, bag limits, seasonal closures and gear restrictions. This section also contains information on the process by which the above measures may get changed in the future.

Table 3: Outline of the Queen Conch Fishery Management Plan for Puerto Rico and US Virgin Islands

> Executive Summary
>
> Definitions
>
> Introduction
>
> Description of resource
>
> Description of fishery
>
> Problems in the fishery
>
> Management objectives
>
> Management measures and alternatives
>
> Recommendations to local governments and other agencies
>
> Related management jurisdictions, laws and policies

The final section of the plan includes information on all the USA federal and local (USA Caribbean) legislation and policies that impinge on the operations of the queen conch fishery. Examples of these are the Federal and Local Endangered Species Acts or the National Environmental Policy Act. For each of these policies or Acts a summary of its relevance to fishery management is provided.

6.4 Western Australia plan for developing new fisheries

Many countries still see fisheries development as one of the pillars of their fisheries policy. In developing countries, the need for creating socio-economic opportunities, generating employment and obtaining hard currency often creates even greater pressure for maintaining this "fishery development" agenda. Although FAO statistics have shown that the prospect for development of new fisheries is small (FAO, 2000), at small scales there will be an on-going need to have procedures in place that ensure the orderly development (or re-development) of new fisheries (Code of Conduct, Paragraph 7.5.4).

Several states in Australia have created management procedures specially designed for this. These procedures, whether in the form of a formal FMP (as it is often done in Queensland) or as a set of general principles, as done in Western Australia, can be a useful guide to how to proceed with developing an FMP for new fisheries in a responsible way (Halmarick, 1999).

Western Australia's guide defines first what constitutes a new fishery:

> ".. a fishery for which there is little or no exploitation, there is potential for development..."

Then it states an important principle that is:

> "...there should be no assumption that the existence of a fish resource will guarantee that commercial access to this resource will be granted."

This clearly establishes that the management authority has the responsibility to define which new resources are to be developed and which not. Whether a new fishery is developed or not should be consistent with the national fisheries policy and should specially consider the possible interaction of this development with existing fisheries.

The next section of the guide then clearly defines the constraints under which development can take place:

- national fisheries policy;
- principles of ecologically sustainable development;
- precautionary principle.

Then the document sets the rights of developers, recognising that these "pioneers" should benefit from developing a new fishery. This should clearly establish what rights may be conferred to those engaging in fishing on fisheries that have not yet been brought under formal management (see Chapter 6 on user rights)

The rest of the document contains sections detailing the process for creation of a new fishery, seeking expressions of interest from prospective participants and establishing the conditions of operations (Table 4). The Western Australian FMP emphasizes that these conditions are to have a limited life of three years, after which an assessment of whether the fishery should continue will be made.

Table 4: Steps required for the development of a new fishery in Western Australia

Expression of interest – opportunity advertised

Ministerial decision – on whether to agree to the development or not

Application – people apply according to guidelines established by management agency

Assessment of applications

Notification of approval/refusal

Implementation – fishing starts

Review and assessment – permits are continued, modified or rescinded

---

7. CONCLUSIONS: HOW DOES A MANAGEMENT PLAN HELP TO ACHIEVE THE MANAGEMENT OBJECTIVES OF A FISHERY?

The previous section showed how plans have to be developed in a way that fits the capacity and needs of each country/fishery. In contrast to the plan for the Australian NPF, the Barbados FMP is a document which purports to describe the management of fisheries for a whole nation, so it includes elements of a national fisheries policy document as well as those elements describing the operational management that are the only ones covered in the NPF FMP. In countries like Barbados, where there is a need to develop both a national fisheries policy at the same time as operational plans for each fishery, it is clearly appropriate to develop an FMP with fishery policy elements in it. Waiting to develop the fishery policy first and the operational FMP later would have deprived the fisheries sector in Barbados of the opportunity to start addressing some of the important management issues that they already face. By contrast the FMP for the NPF or the Queen conch in the USA Caribbean was largely developed to comply with the established policy of the Australian and USA Governments and most of the elements of the plan were already contained in management arrangements that had been developed through many years of management.

For developing countries, which may not have the resources to develop individual FMPs for all existing fisheries, it is unlikely they will be able to develop an FMP for new fisheries. A possible solution is to incorporate, in a special section of the national fisheries management plan, procedures similar to those developed by Western Australia for the development of new fisheries. By doing this the management agency will be able to create a structure through which proper development of new fisheries can occur and avoid the traps of unregulated development.

The lesson is a simple one: FMPs are documents that should first serve to address the most pressing fishery issues faced by each country or fishery. They should do that, however, by at least examining all other issues to ensure that the plan will ultimately encompass all aspects of the fishery to be managed. To achieve this they must be developed to fit within the available legislative and policy framework of each nation or state.

8. REFERENCES

Anon. 1995. Northern Prawn Fishery Management Plan. Attorney-General Department, Australian Government, Canberra. 32 pp.

Anon. 1996. Fishery Management Plan, regulatory impact review and final environmental impact statement for the queen conch resources of Puerto Rico and the United States Virgin Islands. Caribbean Fishery Management Council. 56 pp

Anon. 1997. Barbados Fisheries Management Plan. Schemes for the management and development of fisheries in the waters of Barbados. Fisheries Division, Ministry of Agriculture and Rural Development, Barbados. 67 pp.

FAO. 1997. Fisheries Management. *FAO Technical Guidelines for Responsible Fisheries*, No. 4. FAO, Rome. 82 pp.

FAO. 2000. *The State of World Fisheries and Aquaculture*. FAO, Rome. 142 pp.

Halmarick L. 1999. Developing new fisheries in Western Australia. A guide to applicants for developing fisheries. *Fisheries Management Paper*, **130**. Fisheries Western Australia, Perth, Australia. 40 pp.

8.1 Web resources

The following is a list of www links where copies of FMPs for several fisheries can be downloaded.

Australia

South Australian Government	www.pir.sa.gov.au
Tasmanian Government	www.dpif.tas.gov.au
Western Australian Government	www.wa.gov.au/westfish
Queensland Fish Management Authority	www.qfma.qld.gov.au

United States

Caribbean Fishery Management council	www.caribbeanfmc.com
Mid Atlantic Fishery Management Council	www.mafmc.org
North Pacific Fishery Management Council	www.fakr.noaa.gov
New England Fishery Management Council	www.nefmc.org

Western Pacific Regional Fishery Management Council   www.mar.dfo-mpo.gc.ca

Canada

Canadian Maritimes Region   www.mar.dfo-mpo.gc.ca

GLOSSARY

The terms in this glossary are taken from a number of sources, but particularly from the FAO Technical Guidelines No. 4: Fisheries Management and from the glossary found on the home page of the FAO Fisheries Department (http://www.fao.org/fi/glossary/default.asp). The latter also includes a large number of other fisheries terms.

Bag limit	The number and/or size of a species that a person can legally take in a day or trip.
Biological diversity or biodiversity	The variability among living organisms from all sources including, inter alia, terrestrial, marine and other aquatic ecosystems and the ecological complexes of which they are part; this includes diversity within species, between species and of ecosystems. Diversity indices are measures of richness (the number of species in a system); and to some extent, evenness (variances of species' local abundance). They are therefore indifferent to species substitutions which may, however, reflect ecosystem stresses (such as those due to high fishing intensity).
Biological reference points	A specific type of reference point. A biological reference point indicates a particular biological state of a fishery resource indicator corresponding to a situation considered as desirable (Target reference point, TRP) or undesirable and requiring immediate action (Limit reference point, LRP, and Threshold reference point, ThRP)
Biological resources	These include genetic resources, organisms or parts thereof, populations or any other biotic component of ecosystems with actual or potential use of value for humanity.
Bycatch	Species taken in a fishery targeting on other species or on a different size range of the same species. That part of the bycatch which has no human value is discarded and returned to the sea, usually dead or dying
By-mortality	By-mortality is the mortality of marine organisms from injuries caused by encounters with the fishing gear during the fishing process.
Capital stuffing	The tendency to invest excessively in productive inputs (such as hull, engine, gear). Such investments in fishing capacity are often made to offset regulations to reduce fishing effort.
Catch-per-unit-effort	The quantity of fish caught (in number or in weight) with one standard unit of fishing effort; e.g. number of fish taken per 1000 hooks per day or weight of fish, in tons, taken per hour of trawling. CPUE is often considered an index of fish biomass (or abundance). Sometimes referred to as catch rate.
Co-management	A partnership arrangement in which government and the legitimate interested parties in a fishery share the responsibility and authority for the management of a fishery.
Community-based management	A form of co-management where a central role for management is delegated to a community and where Government would usually have a minor role.
Demersal resources	Species living in close relation with the bottom and depending on it. Example: Cods, Groupers and lobsters are demersal resources. The term "demersal fish" usually refers to the living mode of the adult.

Discards	Are those components of a fish stock (see below) thrown back after capture. Normally, most of the discards can be assumed not to survive.
Efficiency	Obtaining optimal benefits for a given set of inputs, or 'doing the most with what we have'; this can be measured at various levels: the individual fisher or vessel, the fleet, the fishery as a whole, or the coastal region, depending on what level is appropriate. (For example, from the perspective of society as a whole, efficiency may be measured at the scale of what is best for the on-shore economy and relevant coastal communities.)
Exploitation rate	Applied on a fish stock, it is the proportion of the numbers or biomass removed by fishing. A 10% exploitation rate means that 10% of the available stock is being harvested within the time frame considered (per year, per month, etc.). As a measure of fishing pressure, it is proportional to fishing mortality
Fecundity	In general, the potential reproductive capacity of an organism or population expressed in the number of eggs (or offspring) produced during each reproductive cycle. Fecundity usually increases with age.
Fish stock or fish resource	The living resources in the community or population from which catches are taken in a fishery. Use of the term fish stock usually implies that the particular population is more or less isolated reproductively from other stocks of the same species and hence self-sustaining. In a particular fishery, the fish stock may be one or several species of fish but here is also intended to include commercial invertebrates and plants.
Fisheries management organizations or arrangements	These are the institutions responsible for fisheries management, including the formulation of the rules that govern fishing activities. The fishery management organization, and its subsidiary bodies, may also be responsible for all ancillary services, such as the collection of information, its analysis, stock assessment, monitoring, control and surveillance (MCS), consultation with interested parties, application and/or determination of the rules of access to the fishery, and resource allocation.
Fishery	The term fishery can refer to the sum of all fishing activities on a given resource, for example a hake fishery or shrimp fishery. It may also refer to the activities of a single type or style of fishing on a particular resource, for example a beach seine fishery or trawl fishery. The term is used in both senses in this document and, where necessary, its particular application is specified.
Fishing capacity	This is a concept which has not yet been rigorously defined, and there are substantial differences of opinion as to how it should be defined and estimated. However, a working definition is the quantity of fish that can be taken by a fishing unit, for example an individual, community, vessel or fleet, assuming that there is no limitation on the yield from the stock.
Fishing effort	The total amount of fishing activity on the fishing grounds over a given period of time, often expressed for a specific gear type e.g. number of hours trawled per day, number of hooks set per day or number of hauls of a beach seine per day. Fishing effort would frequently be measured as the product of (a) the total time spent fishing, and (b) the amount of fishing gear of a specific type used on the fishing grounds over a given unit of time. When two or more kinds of gear are used, they must be adjusted to some standard type in order to derive and estimate of total fishing effort.

Fishing mortality	A technical term which refers to the proportion of the fish available being removed by fishing in a small unit of time; e.g. a fishing mortality rate of 0.2 implies that approximately 20% of the average population will be removed in a year due to fishing. Fishing mortality can be translated into a yearly exploitation rate (see above) expressed as a percentage, using a mathematical formula.
Fleet	Used broadly in this document to describe the total number of units of any discrete type of fishing activity utilising a specific resource. Hence, for example, a fleet may be all the purse seine vessels in a specific sardine fishery, or all the fishers setting nets from the shore in a tropical multispecies fishery.
Fully exploited	Term used to qualify a stock which is probably neither being overexploited nor underexploited and is producing, on average, close to its Maximum Sustainable Yield. This situation would correspond to fishing at F_{MSY} (in a classical production model relating yield to effort) or F_{max} (in a model relating yield-per-recruit to fishing mortality).
Genetic diversity	The sum of the actual or potential genetic information and variation contained in the genes of living individual organisms, populations or species.
Harvesting strategy	Not to be confused with a management strategy. A harvesting strategy is a plan, under input or output control, for working out how the allowable catch from a stock will be calculated each year e.g. as a constant proportion of the estimated biomass.
High-grading	The discarding of a portion of a vessel's legal catch that could have been sold to have a higher or larger grade of fish that bring higher prices. It may occur in quota and non-quota fisheries.
Interested party or interest group	Refers to any person or group who has a legitimate interest in the conservation and management of the resources being managed. This term is more encompassing than the term stakeholder. Generally speaking, the categories of interested parties will often be the same for many fisheries and should include contrasting interests: commercial/recreational, conservation/exploitation, artisanal/ industrial, fisher/buyer-processor-trader as well as governments (local/State/national). The general public and the consumers could also be considered as interested parties in some circumstances.
Intrinsic rate of increase	The proportional rate of increase of a population at very low population numbers or biomass where density dependent effects are negligible. It therefore represents the average maximum proportional growth rate of the population.
Limited entry	A common management tool in which the government issues a limited number of licenses to fish, which creates a use right - the right to participate in the fishery.
Management authority	The legal entity which has been assigned by a State or States with a mandate to perform certain specified management functions in relation to a fishery, or an area (e.g. a coastal zone). Generally used to refer to a state authority, the term may also refer to an international management organisation.

Management institutions	Used here to indicate arrangements and organisations established to perform specific functions and guide interactions in support of fisheries management. In a broader sense can also be used to describe the set of rules that defines a practice.
Management measure	Specific controls applied in the fishery to contribute to achieving the objectives, including some or all of technical measures (gear regulations, closed areas and time closures), input controls, output controls and user rights.
Management right	The right to be involved in managing the fishery.
Management strategy	The strategy adopted by the management authority to reach the operational objectives. It consists of the full set of management measures applied in that fishery.
Marine protected area	A protected marine intertidal or subtidal area, within territorial waters, EEZs or in the high seas, set aside by law or other effective means, together with its overlying water and associated flora, fauna, historical and cultural features. It provides degrees of preservation and protection for important marine biodiversity and resources; a particular habitat (e.g. a mangrove or a reef) or species, or sub-population (e.g. spawners or juveniles) depending on the degree of use permitted. The use of MPAs (for scientific, educational, recreational, extractive and other purposes including fishing) is strictly regulated and could be prohibited.
Maximum sustainable yield (MSY)	The highest theoretical equilibrium yield that can be continuously taken (on average) from a stock under existing (average) environmental conditions without affecting significantly the reproduction process.
Mortality	The number of deaths in a given period. In fisheries these are divided into those resulting directly from fisheries and those arising from other, 'natural' causes. See also Fishing mortality and Natural mortality.
Natural mortality	A technical term which refers to the proportion of the fish population dying by any causes other than fishing. As with fishing mortality, can be translated into a yearly natural mortality rate expressed as a percentage, using a mathematical formula. See also Fishing mortality.
Non-Governmental Organisation	Any organisation that is not a part of federal, provincial, territorial, or municipal government Usually refers to non-profit organisations involved in development activities.
Objective or Operational objective	A target that is actively sought and provides a direction for management action. For example, achieving a specified income for individual fishers is one possible economic objective of fisheries management.
Open access	A condition of a fishery in which anyone who wishes to fish may do so.
Operational management	Also known as tactical management, involves direct management that affects the fishing process directly, relating to implementation of the management plan and achievement of objectives, including decisions on and implementation of management measures, and monitoring control and surveillance.

Over-exploited	Exploited beyond that limit which is believed to be sustainable in the long term and beyond which there is an undesirably high risk of stock depletion and collapse. The limit may be expressed, for example, in terms of a minimum biomass or a maximum fishing mortality, beyond which the resource would be considered to be over-exploited.
Pelagic resources	Species that spend most of their life swimming in the water column with little contact with or dependency on the bottom. Usually refers to the adult stage of a species
Performance indicator	A specific state, or variable, which can be monitored in a system e.g. a fishery to give a measure of the state of the system at any given time. In fisheries management, each performance indicator would be linked to one or more reference points and used to track the state of the fishery in relation to those reference points.
Productivity	Relates to the birth, growth and death rates of a stock. A highly productive stock is characterised by high birth, growth and mortality rates, and as a consequence, a high turn-over and production to biomass ratio (P/B). Such stocks can usually sustain higher exploitation rates and, if depleted, could recover more rapidly than comparatively less productive stocks.
Property rights	A legal right or interest in respect to a specific property. A type of resource ownership by an individual (individual right) a group (communal right), or the state (state property).
Quota	A share of the Total Allowable Catch (TAC) allocated to an operating unit such as a country, a community, a vessel, a company or an individual fisherman (individual quota) depending on the system of allocation. Quotas may or may not be transferable, inheritable, and tradable. While generally used to allocate total allowable catch, quotas could be used also to allocate fishing effort or biomass.
Recruits	The new age group of the population entering the exploited component of the stock for the first time, or young fish growing or otherwise entering that exploitable component.
Recruitment	The number of fish (recruits) added to the exploitable stock, in the fishing area, each year, through a process of growth (i.e. the fish grows to a size where it becomes catchable) or migration (i.e. the fish moves into the fishing area).
Reference point	An estimated value derived from an agreed scientific procedure and/or an agreed model which corresponds to a state of the resource and/or of the fishery and can be used as a guide for fisheries management. Some reference points are general and applicable to many fish stocks, others should be stock-specific. See also Biological reference point.
Rights-based management	A fisheries management regime in which access to the fishery is controlled by use rights which may include not only the right to fish, but also specify any or all of: how the fishing may be conducted (e.g. the vessel and gear); where they may fish; when they may fish; and how much fish they may catch.
Species assemblage	The term used to describe the collection of species making up any co-occurring community of organisms in a given habitat or fishing ground.
Stakeholder	See Interested party.

Stochastic	Random; involving a random variable (e.g. a stochastic process). Involving chance or probability (syn: probabilistic) (*WWW Webster Dictionary*)
Stock	A group of individuals in a species occupying a well defined spatial range independent of other stocks of the same species. Random dispersal and directed migrations due to seasonal or reproductive activity can occur. Such a group can be regarded as an entity for management or assessment purposes. Some species form a single stock (e.g. southern bluefin tuna) while others are composed of several stocks (e.g. albacore tuna in the Pacific Ocean comprises separate Northern and Southern stocks). The impact of fishing on a species cannot be fully determined without knowledge of this stock structure.
Strategic management	Management of the fishery's overall objectives and policy.
Sustainable use	The use of components of biological diversity in a way and at a rate that does not lead to the long-term decline of biological diversity, thereby maintaining its potential to meet the needs and aspirations of present and future generations.
Target species	Those species that are primarily sought by the fishermen in a particular fishery. The subject of directed fishing effort in a fishery. There may be primary as well as secondary target species
Territorial use rights in fishing (TURFs)	Also Customary Marine Tenure (CMT) - fishery management methods that assign rights to individuals and/or groups to fish in certain locations, generally, although not necessarily, based on long-standing tradition ('customary usage').
Total allowable catch (TAC)	The TAC is the total catch allowed to be taken from a resource in a specified period (usually a year), as defined in the management plan. The TAC may be allocated to the stakeholders in the form of quotas as specific quantities or proportions.
Traditional ecological knowledge	The local knowledge held by a group of indigenous people and passed from generation to generation on the nature and functioning of the ecosystem.
Transboundary stock	Stocks of fish that migrate across international boundaries or, in the case of the United States, across the boundaries between states or Fishery Management Council areas of control.
Trip limit	The right of a specific fisher or vessel to take a certain catch on each fishing trip.
Use rights	The rights held by fishers or fishing communities to use the fishery resources.
Yield	The amount of biomass or the number of units that can be harvested currently in a fishery without compromising the ability of the population/ecosystem to regenerate itself.

AUTHORS' ADDRESSES AND SHORT BIOGRAPHIES

PER ERIK BERGH

Present Position: Adviser to the Permanent Secretary, Ministry of Fisheries and Marine Resources, Namibia

Contact Address
91 Alresford Road
Winchester
Hampshire, SO23 0JZ
United Kingdom

e-mail: daviesbergh@yahoo.co.uk or bigfish@iafrica.com.na
Tel: +264 811242703
Fax: +264 61 223334,

Per Erik Bergh is presently a fisheries management adviser in the Ministry of Fisheries and Marine Resources in Namibia. He specialises in the design, implementation and development of fisheries MCS systems. He has experience from a wide range of situations and countries covering large-scale high technological solutions through to small-scale simplistic approaches to MCS. Mr. Bergh's expertise is based on a background in the Royal Norwegian Navy and Coastguard followed by over ten years of international work in the fields of fisheries and maritime development and collaboration.

ÅSMUND BJORDAL

Department of Marine Resources
Institute of Marine Research
P.O.Box 1870
N-5817
Bergen
Norway

e-mail: aasmund.bjordal@imr.no
Tel: +47 55 23 86 90
Fax: +47 55 23 86 87

Asmund Bjordal is Research Director of the Department of Marine Resources at the Institute of Marine Research, Bergen, Norway. His main responsibilities fall within the areas of fish stock assessment and management advice as well as research and development work on responsible fishing methods. He qualified as a Fishing Master in 1973 and was actively involved in fishing over a number of years. He completed his scientific training at universities in Norway and the USA. The main research areas in which he has been involved include fish behaviour studies for the development of responsible fishing methods, particularly for stationary fishery gears, gear selectivity, and several areas related to aquaculture.

ANTHONY CHARLES

Management Science / Environmental Studies
Saint Mary's University
Halifax
Nova Scotia
Canada B3H3C3

e-mail: Tony.Charles@StMarys.ca
Tel: +1 902 420-5732
Fax: +1 902 496-8101
Web: http://husky1.stmarys.ca/~charles/

Dr. Anthony Charles is a Pew Fellow in Marine Conservation, and a professor of Management Science and Environmental Studies at Saint Mary's University. His work in fisheries, aquaculture and coastal management combines interdisciplinary teaching and research – on themes of sustainability, socioeconomics, management and policy – with practical experience in both developed and developing nations. He is the author of over seventy publications, including the recent Blackwell Science book Sustainable Fishery Systems.

KEVERN L. COCHRANE

Senior Fishery Resources Officer
Fishery Resources Division
FAO
via delle Terme di Caracalla
Rome 00100
Italy

e-mail: Kevern.Cochrane@fao.org
Tel: +39 06570 56109
Fax: +39 06570 53020

Kevern Cochrane works in the Fishery Resources Division of the UN Food and Agriculture Organization. His responsibilities include assistance in implementation of the FAO Code of Conduct for Responsible Fisheries and the provision of technical support to FAO activities in the Caribbean area and the south east Atlantic. He studied in Zimbabwe and South Africa and worked for a number of years on freshwater fisheries in both countries. Thereafter he moved to Marine and Coastal Management in the Department of Environmental Affairs and Tourism in South Africa where he was head of the Stock Assessment Division until his move to FAO in 1995.

SANDY DAVIES

Present Position: Fisheries Adviser, Marine Fisheries and Resources Sector, Southern African Development Community (SADC)

Contact Address
91 Alresford Road
Winchester
Hampshire
SO23 0JZ
UK

e-mail: daviesbergh@yahoo.co.uk
Tel: +264 (0)81 127 0404
Fax: +264 (0)61 223334

Sandy Davies is a Fisheries Adviser with the Southern African Development Community (SADC) Marine Fisheries and Resources Sector. She began her career as a fisheries scientist working with the marine fisheries of the South Atlantic Ocean before moving to African fresh water and marine fisheries management. She has spent time working with Monitoring Control and Surveillance (MCS) systems in various countries and in particular in developing an integrated approach to MCS within fisheries management. For the past three years she has been working with the Southern African Development Community co-ordinating co-operation and development within the marine fisheries sector.

DAVID DIE

Research Associate Professor
Marine Biology and Fisheries Division
Rosentiel School of Marine and Atmospheric Science
University of Miami
4600 Rickenbacker Causeway
Miami, Florida, 33149,
United States of America

e-mail: ddie@rsmas.miami.edu
Tel: +1 305 361-4022
Fax: +1 305 361-4457
Web: http://www.rsmas.miami.edu/divs/mbf/

Dr. David Die holds a BSc in Marine Biology from the Universidad de La Laguna and a PhD in fish population dynamics from the University of Miami. He has worked as a fishery scientist for the Queensland Department of Primary Industries, the Fishery Resources Division of the FAO Fisheries Department and the Australian CSIRO. Dr. Die is currently a research associate professor at the University of Miami. He has also worked extensively as an FAO consultant on the assessment and management of the shrimp and groundfish fisheries of the Brazil-Guianas shelf.

STEPHEN J. HALL

Chief Executive Officer
Australian Institute of Marine Science
PMB No 3 Townsville MC Qld 4810
Australia

e-mail: s.hall@aims.gov.au
Tel: +61 7 4753 4490
Fax: +61 7 4753 4386

Stephen Hall is CEO of the Australian Institute of Marine Science. He has published extensively on the structure and functioning of marine ecological systems, focussing especially on the effects of natural and human disturbance. This work has recently culminated in a book on the global effects of fishing on marine communties and ecosytems. He is a past chairman of the International Council for the Exploration of the Seas (ICES) Working Group on the Ecosystem Effects of Fishing Activities, was a member of a US National Research Council Panel on the Effects of Trawling and is a recipient of a Pew Fellowship in Marine Conservation.

EVELYN PINKERTON

Associate Professor
Simon Fraser University
888 University Drive
Burnaby
BC V5A 156
Canada

e-mail: epinkert@sfu.ca
Tel: +1 604 291-4912
Fax: +1 604 291-4968

Dr. Pinkerton is a maritime anthropologist specializing in common property theory, with particular attention to the role communities play in the management of adjacent renewable natural resources. She has been instrumental in developing the theory and practice of power-sharing and stewardship through cooperative management agreements. Dr. Pinkerton has conducted field research in fishing communities in British Columbia, Nova Scotia, Washington State, and Alaska. She is currently researching the impact of co-management arrangements on management agencies in the states of Washington and Alaska. She is/has served on numerous local, national, and international boards, panels, advisory committees, and ad hoc think tanks including the International Association for the Study of Common Property, the Beijer Institute of Ecological Economics, the National Academy of Science, and the Canadian Marine Fisheries Panel of the Canadian Global Change Program.

JOHN GEORGE POPE OBE

Professor John G. Pope
NRC (Europe) Ltd
The Old Rectory
Burgh St Peter
Norfolk NR34 0BT, England
United Kingdom

e-mail: popejg@aol.com
Tel: +44 1502 677377
Fax: +44 1502 677377

John Pope has worked in fisheries science and management since 1970. For 28 years he worked at the Lowestoft Laboratory of the UK Ministry of Agriculture Fisheries and Food. (Latterly the CEFAS Agency) until taking early retirement at the end of 1997. In the latter years of his service there he held a senior scientific position and provided policy advice to MAFF administrators on most significant fisheries legislation and on MAFF responses to marine biodiversity issues. Pope is currently Director of NRC(Europe) Ltd. and is a visiting Professor of Fisheries Science at the University of Tromsø, Norway.

He is active in the development of fisheries conservation methodology and is credited with the development of several standard methods. More recently, he has helped to develop understanding of the ecosystem effects of fishing and of the application of the precautionary approach to fishing. He has over 50 publications in refereed journals and conference reports to his name. He has also been very heavily involved with the preparation of reports of international working groups and committees and has worked with FAO over many years. Prof. Pope has chaired a number of international working groups and committees of the International Council for the Exploration of the Sea (ICES) and other international organisations. In latter years he acted as UK delegate to ICES and was elected as a Vice president and Bureau Member of ICES in 1995. Post retirement from MAFF he has continued to serve on a number of International Committees and study groups and is a member of several fisheries management bodies.